Jean B. Soper / Martin P. L

Statistics with Lotus 1-2-3

2nd Edition

Chartwell Bratt   Studentlitteratur

British Library Cataloguing in Publication Data
Soper, Jean B.
    Statistics with Lotus 1-2-3. - 2nd ed.
    1. Mathematics. Applications of microcomputer systems.
    Spreadsheet packages. 2. Microcomputer systems.
    Spreadsheet packages: Lotus 123
    I. Title II. Lee, Martin P.
    510'.28'5536

ISBN 0-86238-244-0

All rights reserved. No part of this publication may be reproduced or transmitted in any form or by any means, electronic or mechanical, including photocopying, recording, or any information storage and retrieval system, without permission in writing from the publisher.

© Jean B Soper, Martin P Lee and Chartwell-Bratt Ltd, 1990

Chartwell-Bratt (Publishing and Training) Ltd
ISBN 0-86238-244-0

Printed in Sweden,
Studentlitteratur, Lund
ISBN 91-44-27062-3

    1 2 3 4 5 6 7 8 9 10 | 1994 93 92 91 90

Contents

Preface to First edition vii
Preface to Second edition ix

Chapter 1: Introduction 1

1.1 Statistics and 1-2-3 1
1.2 Getting Started 3
1.3 Entering Information 6
1.4 File Save 10
1.5 Summary 11

Chapter 2: Exploratory Data Analysis 12

2.1   Introduction 12
2.2   Command Menus 13
2.3   Pie Charts 16
2.4   Simple Bar Charts 18
2.5   XY Graphs 20
2.6   Multiple Bar Charts 21
2.7   Stacked Bar Charts 23
2.8   Line Charts 24
2.9   Graph Options 25
2.10 Summary 27

Chapter 3: Measures of Position 28

3.1 Introduction 28
3.2 Labels and Lines 29
3.3 Overwriting and Pre-Sort 32
3.4 Sorting Data 34
3.5 Minimum, Maximum, Median, Q1 and Q3 36
3.6 Summary 40

Chapter 4: Mean and Standard Deviation 41

4.1 Introduction 41
4.2 Formatting Values 42

4.3	Adjusting Column Width	45
4.4	The Arithmetic Mean	47
4.5	The Standard Deviation	50
4.6	"What If?" Calculations	54
4.7	Summary	55

Chapter 5: Frequency Distributions 57

5.1	Introduction	57
5.2	Blanking Spreadsheet Cells	58
5.3	Forming Frequency Distributions	61
5.4	Relative Frequencies	63
5.5	Graphs of Distributions	66
5.6	Summary	67

Chapter 6: Frequencies and Calculations 68

6.1	Introduction	68
6.2	Mean of Frequency Distribution	68
6.3	Standard Deviation Calculation	71
6.4	Printing Spreadsheets	72
6.5	Summary	74

Chapter 7: Probability 75

7.1	Measuring Probability	75
7.2	Database Facilities	77
7.3	Counting Data Records	79
7.4	Probability Rules	81
7.5	Summary	87

Chapter 8: Probability Distributions 88

8.1	Introduction	88
8.2	Binomial Distribution	89
8.3	The Normal Distribution	95
8.4	Continuous Distributions Probabilities	99
8.5	t Values	101
8.6	Z Values	101
8.7	Chi-squared Values	101
8.8	Summary	101

Chapter 9: Decision Analysis 103

9.1	Introduction	103
9.2	Expected Payoff	103
9.3	Posterior Expected Payoff	106
9.4	Summary	109

Chapter 10: Sampling and Inference 110

10.1	Introduction	110
10.2	Extending our Database	112
10.3	Population Summary Values	114
10.4	Simple Random Sampling	114
10.5	Confidence Limits	116
10.6	Sample Replication	117
10.7	Sample Means Distribution	119
10.8	Summary	120

Chapter 11: Crosstabulation of Data 122

11.1	Introduction	122
11.2	Database	123
11.3	Data Tables	123
11.4	Chi-squared Test	125
11.5	Summary	127

Chapter 12: Regression and Correlation 128

12.1	Introduction	128
12.2	Scatter Diagrams	129
12.3	Regression Calculation	131
12.4	Correlation Coefficient	136
12.5	Prediction	137
12.6	Residuals and Standard Errors	140
12.7	Least Squares Property	144
12.8	Influence of Outliers	144
12.9	Summary	145

Chapter 13: Analysis of Variance 147

13.1	Introduction	147
13.2	One Way Analysis of Variance	147
13.3	Two Way Analysis of Variance	151
13.4	Summary	153

Chapter 14: Time Series Analysis 154

14.1	Introduction	154
14.2	Graphs	154
14.3	Moving Means	156
14.4	Exponential Smoothing	158
14.5	Seasonal Analysis	159
14.6	The Additive Model	160
14.7	The Multiplicative Model	163
14.8	Summary	166

Chapter 15: Financial Calculations 167

15.1	Introduction	167
15.2	Compound Interest	167
15.3	Net Present Value	169
15.4	Regular Payments	172
15.5	Summary	177

Chapter 16: Transformations 178

16.1	Introduction	178
16.2	Positively Skew Distributions	179
16.3	Exponential Growth	181
16.4	Double Log Transformations	183
16.5	Summary	184

Chapter 17: Linear Programming 186

17.1	Introduction	186
17.2	Setting up the Problem	186
17.3	Graphical Solution	188
17.4	Summary	191

Chapter 18: Multiple Regression & Matrices 192

18.1	Introduction	192
18.2	Lotus 1-2-3 Release 2	193
18.3	Multiple Regression	195
18.4	The Variance-Covariance Matrix	198
18.5	Retaining Values in Calculations	200
18.6	Summary	201

Chapter 19: Past, Present and Future 205

19.1	Introduction	205
19.2	VisiCalc	205
19.3	SuperCalc	207
19.4	MultiPlan	209
19.5	Integrated Software	211
19.6	As-Easy-As	212
19.7	Quattro	212
19.8	Current Applications	213
19.9	Future Applications	213

Bibliography	214
Index	217
The Authors	223

Preface to First Edition

This book shows how elementary statistical and business calculations can be performed using the computer spreadsheet package Lotus 1-2-3. Spreadsheets have increasingly been recognised as a versatile tool in the teaching of quantitative techniques; they have considerable advantages over other methods of performing elementary statistical calculations.

This book presents an outline of the quantitative techniques usually taught on elementary statistics courses in any discipline, together with some of particular interest to Business Studies users. It shows readers how to use the Lotus 1-2-3 spreadsheet as a calculation worksheet, assuming no previous knowledge of the package. You will learn to input your data and to instruct the computer to carry out computations one step at a time. The actual arithmetic is performed by the computer, but you can see what you are doing and easily make alterations if you wish, either to correct mistakes or to investigate the sensitivity of the final result to the initial data. Consistency checking and the demonstration of theoretical results are also possible.

Lotus 1-2-3 was chosen for the examples in this book because it has been the best-selling microcomputer package since 1983 and is now in use in a wide variety of environments. Both the names Lotus and 1-2-3 are trademarks of the Lotus Development Corporation, the American firm who wrote and now sell the software. The name refers to the integration of a spreadsheet (1) with graphics (2) and a database (3). The graphics help you to visualise your data and results, whilst the database allows the establishment of a population from which various samples can be drawn for further analysis. Users of other spreadsheet packages should find the ideas in this book helpful, but will have to adapt the detailed instructions.

Spreadsheets were primarily designed for financial calculations. Their use in teaching statistics has been developed by the authors at Leicester University and this book collects together the course material they have prepared.

Without the willing cooperation of our students this book could not have been written. We are indebted to them all for their problems and questions, and especially to Barbara Chillman, Mark Alflatt and Neil Burgess who read various chapters of the text. We are very grateful for the encouragement and helpful criticism of our colleagues. In particular we would like to mention Sue Jeffreys who read the complete typescript, and Michael Gibson who helped in the preparation of material and read parts of the text. Even with a word processor, a skilled typist is still invaluable, and we would like to record our thanks to Jackie Macklin who speedily and efficiently typed the major part of the text. We are also grateful to Paul Warren for help with Microsoft Word and John Landamore for help with using the IBM Quietwriter.

In a book containing many formulae, tables and figures, it is very easy for errors to creep in. Many have been spotted by those who read the typescript. Those mistakes which remain are, of course, entirely our responsibility. Authors' families are renowned for their forbearance, but this does not prevent us from wishing to express our thanks to them. The reward for Philip, Andrew, David, Christopher, Kathleen, Harry and Stephen is to see their names in print.

Preface to Second Edition

Encouraged by users of the first edition, we have included an extra chapter in this edition to cover the new commands of Lotus 1-2-3 release 2. At the same time, the book has been completely reset, using a modern laser printer. Should you prefer to have the worked examples available on a disc, details of how this may be purchased are given at the end of the book.

Release 2 introduces two important additional features: multiple regression and matrix manipulation which we have covered in our new chapter 18. Elsewhere in the book, since release 2 is compatible with release 1, we have remained with release 1 menus for simplicity. Our old chapter 18, therefore, has become number 19. It has also been enlarged to cover As-Easy-As and Quattro, in addition to VisiCalc, SuperCalc and Multiplan.

We are again very grateful for the encouragement and help of our colleagues and students. In particular we would like to thank Ian Hyland for help with transferring the first edition text from a PC to an Apple Macintosh, Major Martin C. Richley REME for help with Microsoft Word 3, and Morfydd Edwards and Chris Stone for access to a Apple Laserwriter IINT.

1
Introduction

- Spreadsheet calculations
- Screen display
- Keyboard
- Labels and Values
- Altering entries
- File Save

1.1 Statistics and 1-2-3

Statistics play a vital role in society today. Not only pure scientists, but students of social sciences, business studies and medicine now need to understand statistics. Many students, however, find statistical theory difficult to grasp and statistical calculations laborious. If you are learning statistics and have access to a computer with a spreadsheet program, such as 1-2-3, this book will help you. It shows how you can use a spreadsheet as a statistical worksheet, laying out your calculation so that you can see where the numbers have come from. This book gives only a basic outline of the statistical calculations it shows how to perform. It can be used alongside any elementary statistical text, such as Hodge & Seed (1977) or Berenson & Levine (1979), with which our formulae are usually compatible.

All the calculations in this book are shown using the Lotus 1-2-3 computer program, which integrates a spreadsheet (1) with graphics (2) and a database (3). If you have Lotus 1-2-3 you will, of course, also have the reference manual supplied, but this book fully explains the commands required for the statistical calculations shown so that no previous knowledge of the package is required. We assume that you have 1-2-3 configured for your machine, and that you know how to format data discs. Users of other spreadsheets should find the ideas helpful, but will have to adapt the detailed instructions; we note in the text where particular care is needed with this.

An electronic spreadsheet is rather like a very large sheet of paper ruled into rows and columns, combined with a pencil, rubber and calculator. You enter your data together with labels to identify it and explain what you are calculating, entering one item per cell. Then at each stage of the calculation you tell the computer what arithmetic it is to do and it does it, quickly and accurately. Each intermediate result forms a part of the spreadsheet, which is thus gradually built up towards the final result. At each stage in the process you can check that the results are reasonable. Whatever you enter is stored by the computer.

The data and formulae, therefore, are available throughout the calculation for inspection, and, if desired, alteration. When a value or formula is changed the computer will, unless instructed otherwise, immediately recalculate all values calculated from it. This allows you to correct mistakes and to carry out "What if?" calculations to see the effects of changing particular values. Various statistical results can be demonstrated by making such changes, and examples of them will be given in this book.

Because all the calculations are carried out by the computer, an extra stage in the process, such as first calculating the mean of the data, is little trouble. This makes it natural to work from basic, definitional statistical formulae with which you can see exactly what is being calculated. When working with a spreadsheet you will not need the calculation formulae which textbooks give to ease the burden of hand calculations. Sometimes, however, it may be helpful to use the spreadsheet to study the equivalence of alternative formulae and to compare the accuracy of their results.

Figures are often more meaningful if they are displayed graphically. Just a few keystrokes with 1-2-3 can turn any set of figures in the spreadsheet into a pictorial display. If you change the values in the spreadsheet 1-2-3 automatically changes the graph to correspond, so that the effect of such alterations can be seen both numerically and pictorially.

A sample survey often collects information on a number of items from each respondent. This data can be entered into the spreadsheet using one row for each respondent and particular columns for the responses to particular questions. The database facilities of 1-2-3 enable you to sort and extract from this data as you wish.

1-2-3 is different from the computer programs generally marketed as statistical software. The latter allow users to perform standard statistical tests, but do not require any understanding of the calculation process. A spreadsheet, however, is a general purpose tool which, although originally designed for financial calculations, is useful whenever tables need to be manipulated. With 1-2-3 the user works through the calculation one step at a time and so gains insight into it, while having the tedious arithmetic performed by computer.

The examples in this book mostly use very small data sets, but the data capacity of 1-2-3 is very large. The spreadsheet contains 2048 rows and 256 columns, potentially enabling it to hold over half a million data values. For small data sets 1-2-3 has certain advantages over a hand calculator. A spreadsheet stores and displays both the data and each intermediate result in the calculation, thus allowing easy detection and correction of mistakes, and the carrying out of "What if?" calculations.

The graphical facilities of 1-2-3 can give immediate visual impact to the data and to results calculated from it. For large data sets 1-2-3 provides a convenient method of computer input, the facility of easily performing simple calculations, and the possibility of outputting the data ready to input to another computer package for more complex calculations.

A spreadsheet is a general purpose tool, useful whenever tables need to be manipulated. This book shows how it can be used for many elementary statistical calculations, but in fact spreadsheets were originally designed for financial modelling. They can also be used to evaluate mathematical series and to solve trigonometrical problems, or simply to interrogate a table which has been set up. With iterative or growth processes, such as compound interest, population explosion, radioactive decay and predator-prey simulations, the spreadsheet user can watch the successive steps towards the final result.

1.2 Getting Started

- **Discs, screen and keyboard**
- **Function key [1/HELP]**

Before you can use a piece of software, that is a computer program such as 1-2-3, you need to familiarise yourself with the hardware, the actual microcomputer, which you are going to use. Most microcomputers consist of three main parts: the system box, which includes the floppy disc drives; the keyboard for inputting data and commands into the microcomputer; and the display screen for outputting results. To get started with 1-2-3 we shall first see how to use the disc drives, then consider the screen and lastly the keyboard.

The 1-2-3 **System Disc** should be gently inserted into the left-hand or bottom ('A') disc drive of your microcomputer, whilst a suitably formatted **data disc** should be placed into the right-hand or top ('B') drive. Some microcomputers require small doors or levers to be closed once the discs have been inserted. The power supply should be switched on and, after a short time, prompts may appear for the date and time. If in doubt about what to type in response, just press the large key on the right-hand side of the keyboard. Some machines call this the [RETURN] key, others the [ENTER] key, whilst yet others label it with a reverse **L**-shaped arrow. We shall refer to it as [RETURN] throughout.

The Lotus Access System menu should appear with the first option, 1-2-3 itself, highlighted as in table 1.1.

Table 1.1 The Lotus Access System Menu

```
Lotus Access System   V.1A    (C)1983 Lotus Development Corp.
----------------------------------------------------------------
 1-2-3    File-Manager    PrintGraph    Translate    Exit
Enter 1-2-3 -- Lotus Spreadsheet/Graphics/Database program
================================================================
```

Press the [RETURN] key to select 1-2-3. The screen will then clear and, after a time, the 1-2-3 start-up screen will appear. Press any key to bring the spreadsheet onto the screen as shown in table 1.2.

Normally the **screen** consists of two parts: the **control panel** occupies the top three lines at the top of the screen and the worksheet proper fills the rest of the screen. The top line is the **status line**, which indicates in what cell the pointer is currently located; for example **A1** means that the **cell pointer** is in the top left-hand cell of the spreadsheet. It also gives the current contents of that cell and some other information about what the spreadsheet is expecting next. Below this is the **entry line** where any typing is displayed with the position for the next character being indicated by a **cursor**. The [BACKSPACE] key can be used here for immediate corrections. It is also used for a **menu** or selection of **commands**. The third line is the **explanation line**. It appears when a menu is active, to explain the various options available next from the option at present highlighted in the menu line immediately above.

The worksheet proper consists initially of a blank screen with column letters across the top and row numbers down the left-hand side. Immediately after starting 1-2-3, the **pointer** will be highlighting A1, the top left hand position in the spreadsheet, as the **current cell**.

Table 1.2 The Spreadsheet

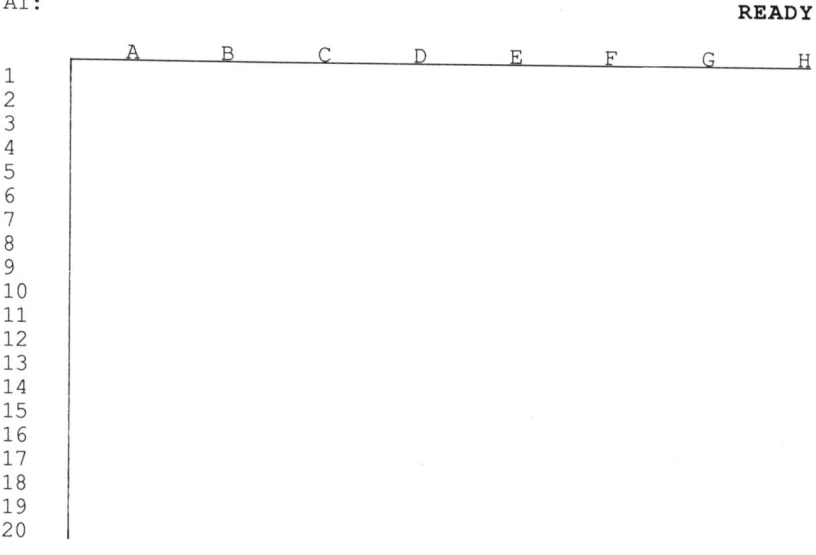

Each make of microcomputer has a slightly different **keyboard**; you will find a diagram in your 1-2-3 manual for your particular machine. Most consist of three main groups of keys: the typewriter keyboard proper; the function keys; and the pointer-movement keys. This last group, which we shall refer to as **arrow** keys, is particulary important in 1-2-3. Some keyboards have other groups of keys such as a numeric keypad, whilst others have combined numeric and arrow key groups.

The typewriter part of the keyboard has the letters **A** to **Z**, the digits **0** to **9**, various symbols (including punctuation) and various special purpose keys. The letter keys will normally generate the small ('lower case') letters **a** to **z** when pressed. To get capital ('upper case') letters **A** to **Z** it is necessary either to press the [CAPS] key, following which all letters will be in upper case, or to hold down one of the two [SHIFT] keys whilst pressing a letter key for the occasional upper case letter. If you've selected upper case letters by pressing the [CAPS] key 1-2-3 will display the message CAPS in the bottom right hand corner of the screen as a reminder. To get back to lower case letters just press the [CAPS] key again. The digit keys, arrayed across the top of the letter keys, normally generate numbers when pressed. To get the symbols on those keys it is necessary to hold down one of the two [SHIFT] keys whilst pressing the relevant digit key.

The special purpose keys include the [RETURN], [SHIFT], and [CAPS] keys already mentioned and some others such as the [ESC] and

[BACKSPACE] keys. The [RETURN] key is always the most important key on any keyboard since it is the one that gets things done. You usually press it to indicate that you have finished typing and you want the computer to DO something. The [ESC] key is the opposite of the [RETURN] key. It tells 1-2-3 to ignore your last instruction and go back a step. However some things can't be undone quite so easily, as we shall see! The [BACKSPACE] key rubs out the character just typed, and is used to correct typing mistakes provided that you have not yet pressed the [RETURN] key.

1-2-3 uses ten **function** (or Programmable Function 'PF') **keys** for particular tasks. On some machines they are set out down the left of the keyboard, whilst on others they are across the top. They are usually numbered 1 to 10, but we shall refer to them by name as well as by number. The most generally useful of the function keys is [**1/HELP**], which calls up the help system. This will display information, taken from the disc, which changes depending on what stage you have reached. If you have the time, try pressing [1/HELP] and following the instructions. Remember, you may press this key in any circumstance for help on your current situation. Press the [ESC] key to return to the spreadsheet.

The four arrow keys, which we shall call [UP], [DOWN], [LEFT] and [RIGHT], are used to move the pointer one cell at a time around the spreadsheet according to the direction of the arrow. We shall see that when entering data it is not necessary to press the [RETURN] key before moving to the next cell by means of an arrow key: the arrow key then simultaneously performs the two functions of entering the data into the current cell and moving on to the next cell in the direction of the arrow.

1.3 Entering Information

- **Labels**
- **Values**
- **Entering using arrow keys**
- **Correcting typing mistakes**

When something is entered in the spreadsheet it is always placed in the current cell. You must, therefore, first position the cell pointer appropriately by using the arrow keys. You can then enter the type of information you wish.

Labels are needed to title your work and to identify particular values. 1-2-3 recognises a label as beginning with a letter (or special prefix). This may be followed by **any** characters you wish, including spaces. .Numerical data can be entered directly into the spreadsheet in its original form. 1-2-3 recognises a number as a **value**.

We shall now build up our first example. Every spreadsheet, like every other document, needs a title to identify it. We can **enter** a title of any length directly into cell A1. For example, type in EMPLOYEE PERSONAL DATA. What you type appears in the entry line above the spreadsheet. If you make a **mistake** use the [BACKSPACE] key to rub it out and then retype it. You can get capital letters by holding down the [SHIFT] key whilst you type but it's easier to press the [CAPS] key once! Immediately you type in the first letter, in this case **E**, notice that the READY message in the top right hand corner of the control panel changes to LABEL. This shows that 1-2-3 has recognised that it is to receive a piece of text rather than a numerical value. At the end of the title label, press the [RETURN] key to indicate that you have finished. 1-2-3 will place the title into cell A1 and into the top line of the control panel, although in the latter case it will be prefixed by an apostrophe. We shall explain the apostrophe later. If you find a mistake after you have pressed the [RETURN] key it is probably quickest, at this stage, to retype the title again from scratch, and then press the [RETURN] key. It will be entered in cell A1 in place of the previous entry. You may be able to see that your title label is, in fact, wider than cell A1 and has overspilled into B1 and C1. That is permissible so long as we do not want to put something else in these cells. The spreadsheet should now look like table 1.3.

Table 1.3 A Title Label

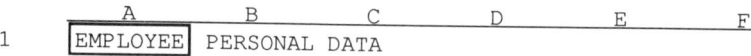

	A	B	C	D	E	F
1	EMPLOYEE	PERSONAL	DATA			

We are going to build a table of information about five people. Each row of the table refers to one person, or case, and is said to constitute a record. Each column contains a different piece of information about the various people. It may be an attribute, such as the person's sex, or a variable, such as his or her age, measured in years.

Use the [DOWN] arrow key twice to move the pointer to cell A3. We are going to place a series of column labels for the attributes or variables across row 3. Type in SURNAME: but at the end do not press the [RETURN] key; instead, if you are happy with the column label, press the [RIGHT] arrow key. This has two effects: it places the label into cell A3 and moves the pointer into B3. Now type in FORENAME:, again followed by the [RIGHT] arrow key. Next type AGE: and the [RIGHT] arrow, and lastly SALARY: followed by the [RIGHT] arrow key. You should now have the cell pointer in cell E3 with four column labels to the left of it. If you find that you have mistakenly entered an item into a cell you intended to leave blank, for example cell A2, for the moment you can replace that entry with a space which will make the cell appear blank. When the message READY is showing and the A2 cell is highlighted press [SPACE] then [RETURN]. Thus a space is entered and the cell appears blank. Spaces can cause problems, so later we shall use the correct way to blank out unwanted cell entries. The spreadsheet should now look like table 1.4.

Table 1.4 Column Labels

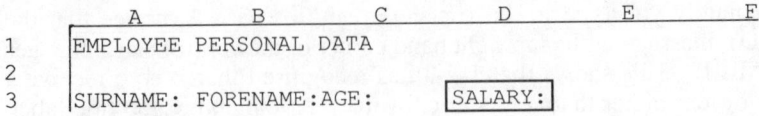

Move the pointer [DOWN] twice and [LEFT] four times to get to cell A5. We are now going to place a series of row labels down column A. Type in Smith followed by the [DOWN] key. Create a column of row labels to give a spreadsheet like table 1.5.

Table 1.5 Row Labels

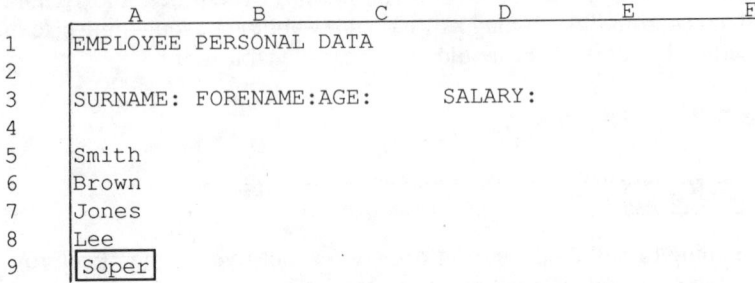

Press the [RIGHT] arrow once and the [UP] arrow a few times to get to cell B5 ready to fill in the FORENAMES in table 1.6.

8

Table 1.6 More Labels

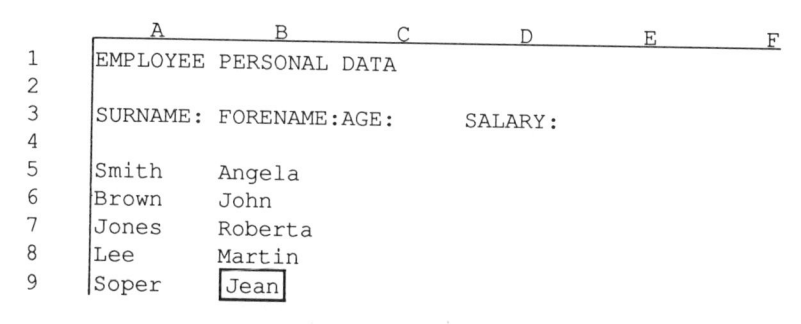

```
      A         B         C         D         E         F
1  EMPLOYEE PERSONAL DATA
2
3  SURNAME:  FORENAME:AGE:       SALARY:
4
5  Smith     Angela
6  Brown     John
7  Jones     Roberta
8  Lee       Martin
9  Soper     Jean
```

Move the pointer to C5 and type in the number 20. Since the first key pressed is a digit rather than a letter 1-2-3 changes the READY message to VALUE rather than LABEL. As with labels, numbers can be entered into the worksheet by pressing either the [RETURN] key or an arrow key. Fill in the rest of the age values to give table 1.7.

Table 1.7 Values of a variable

```
      A         B         C         D         E         F
1  EMPLOYEE PERSONAL DATA
2
3  SURNAME:  FORENAME:AGE:       SALARY:
4
5  Smith     Angela       20
6  Brown     John         61
7  Jones     Roberta      27
8  Lee       Martin       32
9  Soper     Jean         25
```

Move the pointer back up to cell D5 and type in the following data values for SALARY to get table 1.8.

Table 1.8 Values of another variable

```
      A         B         C         D         E         F
1  EMPLOYEE PERSONAL DATA
2
3  SURNAME:  FORENAME:AGE:       SALARY:
4
5  Smith     Angela       20      4800
6  Brown     John         61     15460
7  Jones     Roberta      27      5110
8  Lee       Martin       32     10655
9  Soper     Jean         25      6325
```

1.4 File Save

You have now completed your first spreadsheet but before you turn your computer off you must save it on a suitable data disc in the B drive. **Always save** what you have done, even if it is unfinished and even if you are not proud of your first attempt. It's usually easier to correct mistakes than to start again from scratch. You'll need the spreadsheet for use in the next chapter.

Make sure 1-2-3 is READY, then press the [/] key and the control panel will change to that shown in table 1.9. The READY message will change to MENU, the entry line will consist of the nine commands of the **main menu** and the explanation line will comprise the eight options of the **Worksheet** sub-menu. The word Worksheet will be highlighted by the menu pointer (similar to, but not to be confused with, the cell pointer).

Table 1.9 The Main Menu

```
A2:                                                           MENU
|Worksheet|  Range Copy Move File Print Graph Data Quit
Global,Insert,Delete,ColumnWidth,Erase,Titles,Window,Status
```

You can now select the **File** option either by pressing the [RIGHT] arrow key four times followed by the [RETURN] key or by pressing the **F** key directly. The control panel alters to show the next set of choices as in table 1.10.

Table 1.10 The File Sub-Menu

```
A2:                                                           MENU
|Retrieve|  Save Combine Xtract Erase List Import Directory
Erase the worksheet and read a worksheet file
```

Again to select the **Save** option you can either press the [RIGHT] arrow once followed by the [RETURN] key or simply press the **S** key. The computer responds by asking for a file name. The control panel reads as in table 1.11.

Table 1.11 The File Save Command

```
A2:                                                           MENU
Enter save file name:
```

10

Type in a **filename** consisting of **up to eight** letters or digits, for example EMPDATA1, followed by the [RETURN] key. Your spreadsheet will then be saved as a file on the disc for later retrieval.

When the disc drive has stopped working you can press the [/] key to call up the main menu again and then press **Q** to Quit from 1-2-3. Then press **Y** for Yes to indicate you really do want to quit. You may be returned to the Lotus Access System in which case press **E** for Exit and **Y** for Yes to indicate you do really want to exit! You should get back to A> on the screen, when you can safely remove your disc(s) and then switch off your computer if no-one else is waiting.

1.5 Summary

Spreadsheets offer a different way of doing statistical calculations. You utilise the data storage and computational facilities of a computer, but you set out the calculation so that you can understand exactly how it is performed.

This book will help you learn both about the 1-2-3 spreadsheet and about methods of statistical computation.

2
Exploratory Data Analysis

- **Menus**
- **Ranges**
- **Graphics**
- **Printing graphs**
- **Worksheet Erase**

2.1 Introduction

In the previous chapter we saw the entry of data into the spreadsheet, and now we can move on to explore that data simply by looking at it. A picture is often said to be worth a thousand words; it is also worth a thousand numbers! With a computer we can draw even complicated graphs quite easily and accurately. The conversion of numbers into pictures enables the general form of the data to be seen at a glance whilst allowing specific, exceptional values to stand out. In the previous chapter we also briefly encountered 1-2-3's command menus. We shall use some more commands for graphing and so shall begin with a quick review of the menu system.

2.2 Command Menus

- Selecting commands
- Backing out of selected commands
- Defining ranges

If you press the [/] key you will call up the main menu:

Table 2.1 The Main Menu and Worksheet Sub-Menu

```
A2:                                                          MENU
|Worksheet|  Range Copy Move File Print Graph Data Quit
Global,Insert,Delete,ColumnWidth,Erase,Titles,Window,Status
```

Remember that the entry line now contains the nine commands of the main menu with the first one, Worksheet, highlighted by the menu pointer. The explanation line now consists of the eight commands of the **Worksheet sub-menu**. These are general purpose commands operating on entire rows or columns of the spreadsheet. If you now press the [RIGHT] arrow key the menu pointer will move on to the Range command and the explanation line will carry the eight commands in the Range sub-menu. Pressing the [RIGHT] arrow key repeatedly **highlights** the main menu options in turn and displays the sub-menus corresponding to each.

Table 2.2 The Range Sub-Menu

```
A2:                                                          MENU
Worksheet  |Range|  Copy Move File Print Graph Data Quit
Format,LabelPref,Erase,Name,Justify,Protect,Unprotect,Input
```

The **Range sub-menu** refers to sections of the spreadsheet. A **range** can be a single cell, a line of cells along part of a row or column, or even a rectangular array of cells. The Range sub-menu provides for various operations to be undertaken on such a group of cells. The next choice in the main menu after Range is the Copy command. It is highlighted by pressing the [RIGHT] arrow key:

Table 2.3 The Copy Command

```
A2:                                                          MENU
Worksheet Range  |Copy|  Move File Print Graph Data Quit
Copy a cell or range of cells
```

The **Copy command** is used to copy the contents of a single cell into another single cell or range of cells. It can also copy a range into another range. It copies not just labels and values but formulae as well, adjusting cell addresses automatically. Copying saves much repetition and will be used frequently in later examples. Pressing the [RIGHT] arrow key again gives the Move command:

Table 2.4 The Move Command

```
A2:                                                          MENU
Worksheet Range Copy  Move  File Print Graph Data Quit
Move a cell or range of cells
```

The **Move command** is used to move a single cell into another single cell or to move a range into another range. The spreadsheet does not close up over the cell(s) from which information has been moved nor does it open up over the cell(s) to which information is being moved. This means that great care is needed to ensure that the receiving area is empty. Moving can greatly change the layout of a spreadsheet. Use the [RIGHT] arrow key again to obtain the File sub-menu:

Table 2.5 The File Sub-Menu

```
A2:                                                          MENU
Worksheet Range Copy Move  File  Print Graph Data Quit
Retrieve,Save,Combine,Xtract,Erase,List,Import,Directory
```

The **File sub-menu** provides access to and from the disk. We have already met it when we saved the example spreadsheet in the previous chapter. Pressing the [RIGHT] arrow key again gives the Print command:

Table 2.6 The Print Command

```
A2:                                                          MENU
Worksheet Range Copy Move File  Print  Graph Data Quit
Output a range to the printer or a print file
```

The **Print command** sends part or all of the spreadsheet to the printer (if you have one attached to your computer). Selecting this command in fact gives rise to a sub-menu, the commands in which we shall describe in chapter 6. The Graph command will be previewed next with the [RIGHT] arrow key:

Table 2.7 The Graph Command

```
A2:                                                          MENU
Worksheet Range Copy Move File Print  Graph  Data Quit
Create a graph
```

The **Graph command** is used to generate graphs on the screen and we shall be using it extensively later in this chapter. Pressing the [RIGHT] arrow key again gives the Data sub-menu:

Table 2.8 The Data Sub-Menu

```
A2:                                                        MENU
Worksheet Range Copy Move File Print Graph [Data] Quit
Fill, Table, Sort, Query, Distribution
```

The **Data sub-menu** provides access to 1-2-3's database operations which we shall be using in later chapters. Pressing the [RIGHT] arrow key once more gives the Quit command:

Table 2.9 The Quit Sub-Menu

```
A2:                                                        MENU
Worksheet Range Copy Move File Print Graph Data [Quit]
End 1-2-3 session (Have you saved your work ?)
```

The **Quit command** is used to leave 1-2-3; we met it briefly in the last chapter.

Notice that if you keep pressing the [RIGHT] arrow key you will wrap around back to the first command, whereas repeatedly pressing the [LEFT] arrow key will highlight the command to the left until that too will wrap around the other way. When you are searching for a command you can try all the different possibilities on the entry line in turn to see, on the explanation line beneath, what they would do if selected. To **select** a highlighted command you press the [RETURN] key. When you become a little more familiar with the menu structure you can just type the **first letter** to select a command, as we mentioned at the end of the last chapter.

Notice also that sometimes the explanation line is itself a menu, called a **sub-menu** (e.g. Worksheet, Range, File and Data commands), and sometimes just a sentence (e.g. Copy, Move, Print, Graph and Quit commands). Actually the Print and Graph commands do also lead to sub-menus in spite of the sentence on their explanation lines.

When you select a command which has a sub-menu, the sub-menu is then displayed as a menu with the first item highlighted. If this item gives rise to a further sub-menu, this will now be displayed on the explanation line. Some commands lead to several layers of nested sub-menus.

Sometimes you can enter the wrong sub-menu by mistake. To **back out** of a sub-menu just press the [ESC] key. This can be done repeatedly until READY mode is reached. If, however, you are deeply embroiled in sub-menus of sub-menus you can come right back to READY just by pressing the [BREAK] keys together.

The meaning of the term range was outlined briefly above when it appeared as one of the options in the main menu. Range also appears frequently in the above explanation lines and, indeed, 1-2-3 often operates on a range of cells specified by the user. A **range** is defined by its **start** and **end points**, which are: the cell address for a single cell; the addresses of the first and last cells to be included for a column or row; and the addresses of a pair of diagonally opposite corner cells for a rectangular array. For example, when a new spreadsheet is first brought onto the screen the range of cells visible in the first column is **A1..A20**, and in the first two columns **A1..B20**. You can alternatively define the first column range as **A20..A1**, and the rectangular range as **A20..B1**, **B1..A20** or **B20..A1**.

When 1-2-3 requires you to set a range it offers a suggested range or starting point of a range, based either on a previously defined range or on the position of the cell pointer. You may accept the range offered by pressing the [RETURN] key. Alternatively, you may ignore the range offered and type in the range you require, e.g. **A1..A20**. You need only one [.] between the start and end points of a range; the second dot will be displayed automatically by 1-2-3. Another way is to use the pointer to highlight the desired range of cells. If both the start and end of the range are offered in the second line of the control panel you may only alter the starting point by first using the [BACKSPACE] key to rub out the end point. Now move the cell pointer to the desired starting point of the range, say **A1**. Press the [.] key to anchor the starting point and move the cell pointer again to extend the range as desired. Moving it down to **A20** will highlight the range, shown in the control panel as **A1..A20**. Moving the pointer right to **B20** then shows the range **A1..B20**.

Now that we have reviewed the use of the command menus we can try our first graph, a pie chart.

2.3 Pie Charts

- **Graph Type Pie**

Pie charts are graphs in which a circle is subdivided into segments whose sizes are proportional to the various data values which they represent. To draw a pie chart manually involves a pair of compasses to set out the circle and a protractor to measure out angles for the segments. Since a circle subtends 360°, conversion of raw values into angles also requires some arithmetic. To draw a pie chart on the computer requires but a few keystrokes with the calculation and drawing being completely automatic.

In this section we are going to generate a pie chart from the spreadsheet in the last chapter. Fig. 2.1 shows what the finished graph will look like.

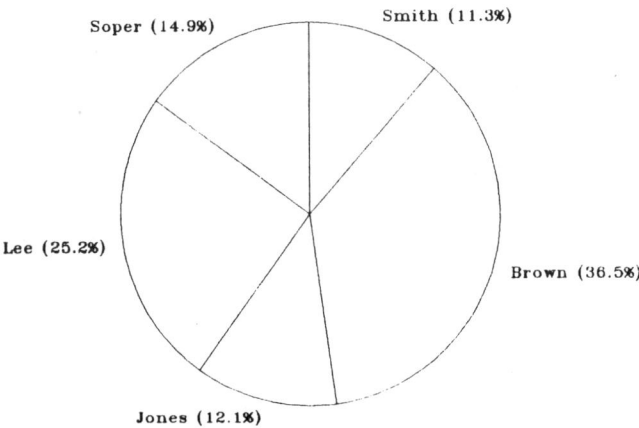

Fig. 2.1 Pie Chart of Individual's Salary

Press the [/] key for the menu, **F** for the **File** menu and **R** to **Retrieve** your first worksheet. You will be offered the filename, e.g. EMPDATA1, and to accept it just press the [RETURN] key. After the spreadsheet has been retrieved from disk and appears on the screen, press the [/] key for the menu again. This time press **G** for **Graph**, obtaining the sub-menu shown in table 2.10.

Table 2.10 The Graph Sub-Menu

```
A2:                                                          MENU
 Type  X A B C D E F Reset View Save Options Name Quit
Set graph type
```

Press the [RETURN] key to select the **Type** of graph required and you will be offered the five types of graph shown in table 2.11.

Table 2.11 The Graph Type Sub-Menu

```
A2:                                                          MENU
 Line  Bar XY Stacked-Bar Pie
Line graph
```

Press **P** for **Pie** chart which will also take you back to the graph sub-menu (Table 2.10). Press **X** to set up the X-range. These are the row labels: they are used to identify the segments of the pie chart. You can either type **A5..A9** or you can move the cell pointer to **A5**, press the [.] key to anchor the start of a range, press the [DOWN] arrow key four times to highlight all the row labels and then press [RETURN]. The first way involves specifying actual cell addresses, whilst the second relies on pointing to the required range of cells. You are now in the graph sub-menu again.

Press **A** to set the first data range. These are the actual values to be graphed and you can again either type **D5..D9** or you can move to **D5**, press the [.] key to pin the start of a range, move down to the end of the range and then press [RETURN]. You can now press **V** to **View** your pie chart!

Notice that 1-2-3 has calculated and plotted percentage values for you. When you have finished looking at your graph press any key to continue.

2.4 Simple Bar Charts

- **Graph Type Bar**
- **Graph Save**

Whereas pie charts are circular, **bar charts** are rectangular with the data values displayed as bars of equal width whose height is proportional to their value. Lotus 1-2-3 provides simple bar charts and stacked bar charts, the latter being useful when one set of values is made up of several other sets.

To change our pie chart into a simple bar chart requires only 3 keystrokes. Press **T** for graph **Type**, **B** for **Bar** chart and then **V** to **View** the result shown in fig. 2.2.

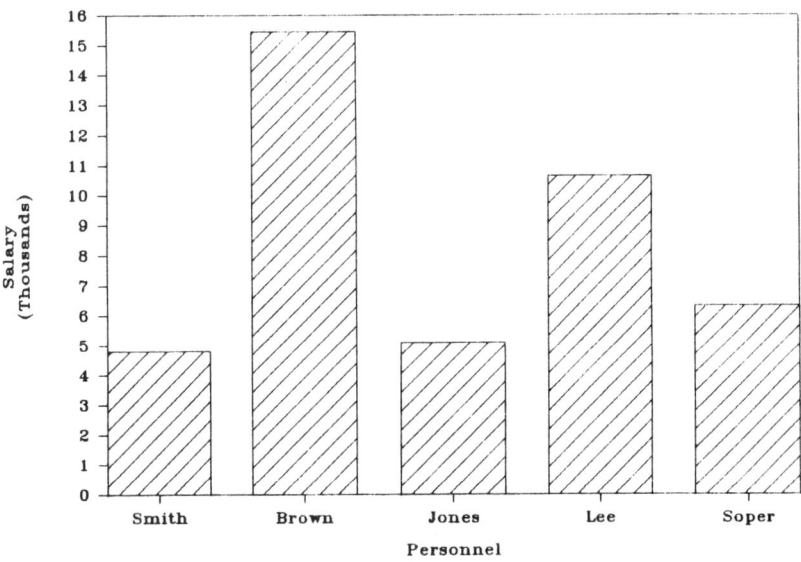

Fig. 2.2 Bar Chart of Individual's Salary

Notice that 1-2-3 has automatically represented the salary figures in thousands.

If you want a hard-copy of your chart on paper then you must press **S** for **Save** and enter a name for the picture, such as EMPBAR1. Notice that this could be the same name as was used for the worksheet itself, since they are different files, as 123 invisibly appends .PIC and .WKS respectively to the file names. Although the worksheet itself looks unchanged it now contains hidden details of the bar chart so it too requires saving. You can press the [ESC] key to escape from the graph menu back to the main menu then press **F** for File and **S** for Save and give the filename EMPDATA2. It is better to give a new name than to use the old one (EMPDATA1), since in case of disc trouble, you will have two, slightly different, copies saved. You can now press **Q** to Quit from 1-2-3, and **Y** for Yes you have saved so you do want to quit. Back now in the Lotus Access System press **P** for the Printgraph program; this will prompt for the relevant floppy disc to be inserted in place of the system disc. For details of the Printgraph program see the Lotus 1-2-3 manual.

2.5 XY Graphs

- **Graph Type XY**
- **Graph Options Format**

XY graphs are used when we have a set of paired measurements on two variables **X** and **Y**. The **X** values are plotted according to a horizontal axis whilst the **Y** values are plotted according to a vertical one. The data pairs are referred to as Cartesian coordinates. XY graphs are the only type of graph in 1-2-3 in which the X range is truly numeric. In all the other graphs the X range, even if numerical, is merely plotted across the bottom of the screen as a range of labels. These graphs are, therefore, only meaningful for equally spaced numerical **X**'s.

To illustrate an XY graph, let us look at the relationship between age and salary in our first worksheet (see Fig. 1.6). If you have printed the previous graph you will need to re-enter 1-2-3 proper, after which you can press [/] for the main menu and select File Retrieve EMPDATA2. When this worksheet has loaded, press [/] to call up the main menu again then select Graph Reset Graph to clear the previous graph settings, choose graph Type XY and set up age as an X range with salary as an A range. Let us plot a scatter diagram, marking the points which represent the pairs of values with isolated symbols. Select the **Options** sub-menu:

Table 2.12   The Graph Options Sub-Menu

```
A2:                                                              MENU
 Legend  Format Titles Grid Scale Color B&W Data-labels Quit
Specify data-range legends
```

Select **Format**, which asks you to choose ranges to format:

Table 2.13   The Graph Options Format Sub-Menu

```
A2:                                                              MENU
 Graph  A B C D E F Quit
Set format for all ranges
```

Select **Graph**, or **A**, which is the only range you have used:

Table 2.14   The Graph Options Format Graph Sub-Menu

```
A2:                                                              MENU
 Lines  Symbols Both Neither
Draw lines between data points
```

Choose **Symbols**, Quit from the Format menu, Quit from the Options menu and then View the resulting scatterplot shown in fig. 2.3.

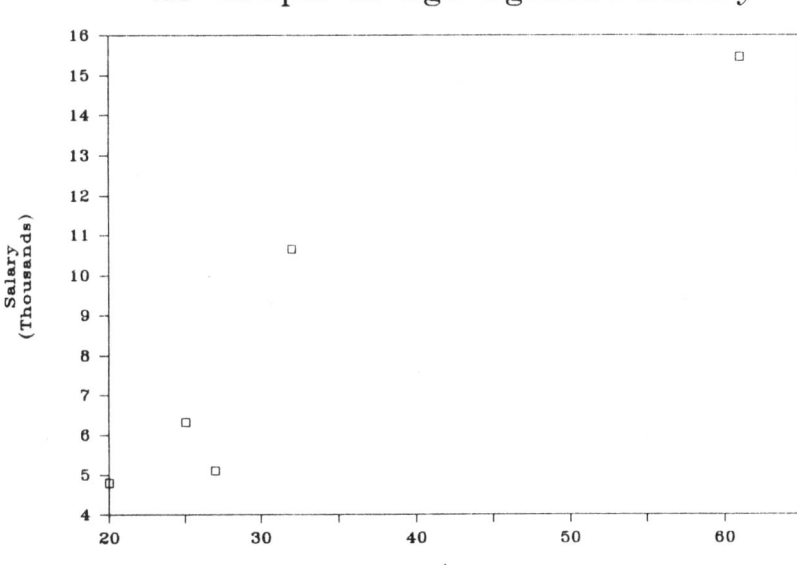

Fig. 2.3 XY Graph of Age against Salary

Notice that although this XY graph has just one set of data on the Y axis (denoted by the A range), XY graphs can have up to six sets of data (A to F ranges) on the Y axis for one set of **X** values.

2.6 Multiple Bar Charts

- **Graph Options Legend**

Bar charts may also have up to six values corresponding to each row label, where the row labels form the X range and the values are in any or all of the A, B, C, D, E or F ranges. To illustrate this we need to modify our spreadsheet. If you still have the worksheet loaded, quit from wherever you are and, once in READY mode, add the information displayed in columns E and F of table 2.15.

Table 2.15 Modified Spreadsheet

	A	B	C	D	E	F
1	EMPLOYEE	PERSONAL	DATA			
2						
3	SURNAME:	FORENAME:	AGE:	SALARY:	BONUS:	COMMISSION:
4						
5	Smith	Angela	20	4800	1000	2120
6	Brown	John	61	15460	1000	7520
7	Jones	Roberta	27	5110	1000	3000
8	Lee	Martin	32	10655	1000	5380
9	Soper	Jean	25	6325	1000	3870

To draw a multiple bar chart from this worksheet, press [/] to call up the main menu and then select Graph Type Bar-chart. Set up the row labels as an X range with successive columns of values of salary, bonus and commission as A, B and C ranges respectively and then View the chart, shown in fig. 2.4. Notice that for each row label there are three bars grouped together. The bars can be identified on the graph as shown by selecting **Options** and then **Legend** and typing appropriate labels.

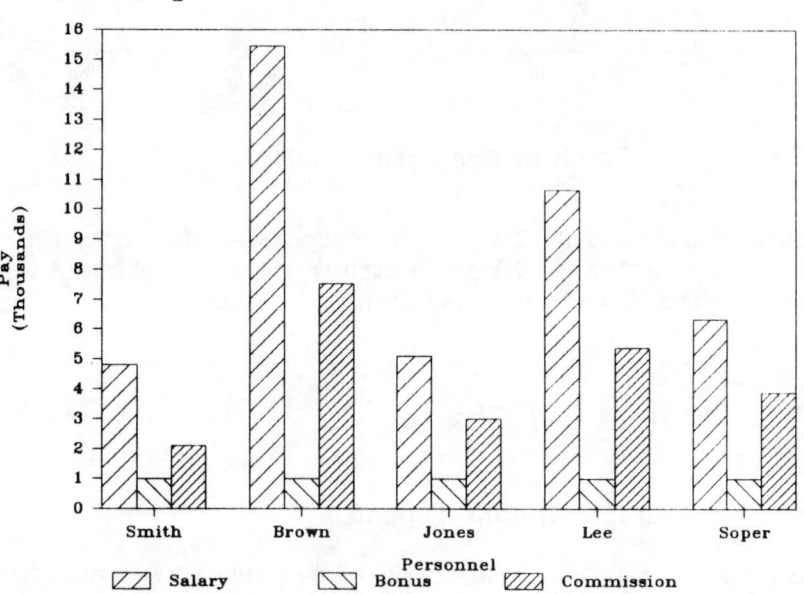

Fig. 2.4 Multiple Bar Chart for Three Variables

2.7 Stacked Bar Charts

- **Graph Type Stacked-Bar**
- **Worksheet Erase**

To generate a **stacked bar** chart also requires more than one datum value for each row, since each bar represents several values which sum together to form a total. They might, for instance, be several percentages which add up to 100% or four quarterly amounts which represent a year. In our example an individual's earnings consist of the sum of salary, bonus and commission, therefore in the Graph menu select Type Stacked-bar and View the result, displayed in fig. 2.5.

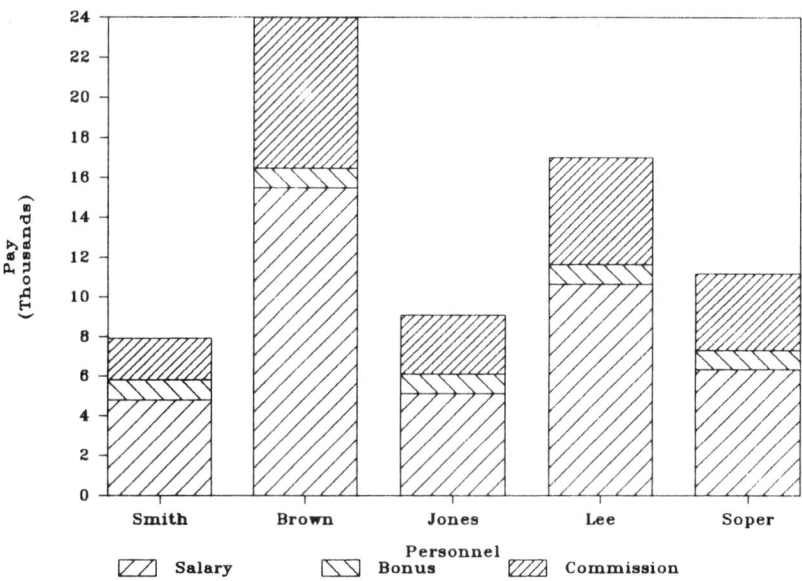

Fig. 2.5 Stacked-Bar Chart

Notice that now the three values for each row label are stacked on top of each other, with the height of each stacked bar proportional to the sum of the relevant values. Stacked bar charts (and Pie charts) by their very nature must not include negative values; if any are present they will be erroneously treated as positive ones.

We can now Save the graph as a picture with a name such as EMPBAR2, press the [ESC] key to retreat back to the main menu and then File Save the worksheet itself as EMPDATA3. Notice that in both cases we have specified a new name rather than replaced the existing version. This provides us with backup in case of disc problems and allows us to retain the original versions for future reference.

If you wish to proceed to create a new worksheet, as in the next section, you should select **Worksheet Erase Yes** to clear the present one, but be careful to save it first. Note also that it is not necessary to erase a worksheet before retrieving a different one from disc since the new one will overwrite the old.

2.8 Line Charts

- **Graph Type Line**

Line charts are like simple bar charts but with data points plotted at the top of the bars and joined together by straight lines. Whereas the row labels in pie and bar charts are usually just nominal categories, in a line chart the row labels are ordinal categories such as equal time periods. Remember that if your X values are not equally spaced you should use an XY graph.

Table 2.16   A New Spreadsheet

```
        A            B           C
1   ANGELA SMITH'S COMMISSION 1985
2
3   QUARTER: COMMISSION:
4
5   QTR 1          250
6   QTR 2          280
7   QTR 3          540
8   QTR 4         1050
```

Enter the information shown in table 2.16. Press [/] for the main menu and then select **Graph Type Line**, set up the four quarters as an X range and the commission values as an A range and then View the result. When you are happy with the graph, File Save the worksheet as COMDATA1.

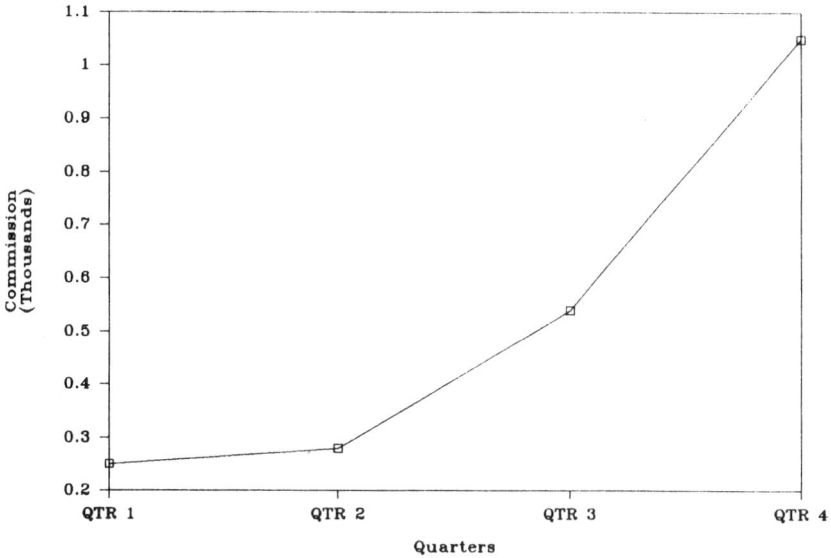

Fig. 2.6 Line Chart

2.9 Graph Options

Lotus 1-2-3's graphics are a powerful feature and are particularly important in statistical work since they can turn a mass of numbers into a coherent picture. However to get maximum benefit from a graph it needs to be laid out carefully. For this we need to look at the Graph Options sub-menu in more detail:

Table 2.17 The Graph Options Sub-Menu

```
A2:                                                          MENU
 Legend  Format Titles Grid Scale Color B&W Data-labels Quit
Specify data-range legends
```

1-2-3 allows us to label the various symbols and/or shadings used on our graphs with a **Legend**, this is particularly useful for stacked bar, line or XY graphs when they have several data ranges. The **Format** option specifies whether lines, symbols, both or neither are drawn on line and XY graphs. 1-2-3 gives us up to four **Titles** on a graph, two at the top and one on each axis. These are essential on all graphs. **Grid** gives horizontal, vertical, both or neither sets of lines across the graph for reference purposes. Because they clutter up a chart, they make it harder to see the overall picture it presents, but they make it easier to read individual values, particularly on the screen, from the chart. Scaling is normally done automatically but sometimes it is useful to change the end-points of the axes with the **Scale** option. The **Color**/black and white switch depends on what sort of monitor your computer supports. The **Data-labels** option allows the superimposition of worksheet labels onto the body of a graph. This can also lead to a cluttered graph, but can be useful, for example, for labelling points on the graph. As you can see Graph Options are very extensive: as you go through them, write down the choices you have made until you are happy with all the posibilities.

To illustrate the combination of the various graph options we have reworked a previous figure. See if you can reproduce it exactly:

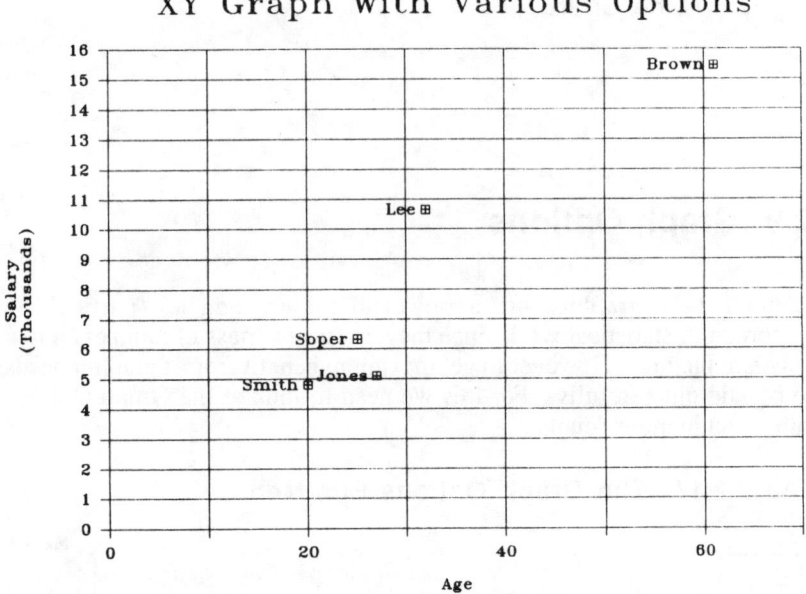

Fig. 2.7 XY Graph with Various Options

2.10 Summary

Lotus 1-2-3 provides a reasonable set of different graphs which greatly aid our understanding of statistics. Apart from the five built-in graph types, other more complicated charts and even pictures can be drawn by manipulating the basic ones. However the real strength of 1-2-3 is the speed and facility with which we can summon up a graph of our numbers with just a handful of keystrokes. This means that we can look at the numbers in our worksheet repeatedly as it develops, checking the consistency of each new manipulation. When we have finished our computations we can spend a little more time to generate really first class graphics for presentation purposes.

This chapter is entitled Exploratory Data Analysis (EDA) and we hope that it will encourage you to **look** at your data before you try any statistical manipulation. Traditionally EDA has concentrated on simple pencil and paper graphical techniques with the minimum amount of arithmetic. However, 1-2-3 gives us rather more sophisticated graphics with no arithmetic to perform at all, and therein lies one of its strengths.

3
Measures of Position

- Label prefixes
- Copying labels
- Data Sort
- Simple formulae
- Pointing at cell addresses
- Range Name

3.1 Introduction

A set of data values can be summarised using five measures of position: the minimum, lower quartile, median, upper quartile and maximum values of the data. To find these summary values it is convenient first to arrange the data in ascending size order. This is easily accomplished in 1-2-3 by using the Data Sort command. The **minimum** and **maximum** values can then be identified as the lowest and highest data values respectively. The **median** is the middle value of the ordered set of data. If there are altogether **n** values and **n** is odd the median, **Md**, is the $(n+1)/2^{th}$ value. If **n** is even we take the median to be halfway between the $n/2^{th}$ and $(n+2)/2^{th}$ values.

The **quartiles Q1** and **Q3** are similar to the median, lying respectively one quarter and three quarters of the way through the ordered data set. The first quartile, **Q1**, then, is the $(n+1)/4^{th}$ value of the distribution, the third quartile, **Q3**, is the $3(n+1)/4^{th}$ value, and we may again interpolate between values if necessary.

In this chapter we shall extend the employee salary data which we had in our first spreadsheet and we shall calculate the various measures of position for that data. You should, then, again retrieve the file EMPDATA1, which is shown in table 1.8.

Perhaps you have wondered whether we could improve the appearance of this spreadsheet, for example by aligning the column titles with the values beneath them. We shall begin by seeing how this can be done, by using **label prefixes** to alter the positions of our labels. Notice that these prefixes can **only** be used to position labels and not values. We shall see in Chapter 4 how to alter the way in which values are displayed. A line can be drawn on a spreadsheet by entering an appropriate label and making use of the Copy command to replicate it, as will be shown later in this chapter.

The **Copy** and **Data** commands which will be introduced in this chapter require you to define **ranges** in the way described in 2.2. Since 1-2-3 will offer a suggested range, you will find it convenient to ensure the cell pointer is at a corner of the range you intend to define **before** you call up the menu to begin the command.

3.2 Labels and Lines

- **Label prefixes**
- **Range Label-Prefix**
- **Worksheet Global Label-Prefix**
- **Copy**

When you have retrieved EMPDATA1, position the cell pointer on a cell containing a label, for example A3. The label which has been entered, **SURNAME:**, will be displayed in the control panel, but it will be prefixed by an apostrophe [']. The apostrophe is an instruction to 1-2-3 to position the label so that it is aligned to the left of the cell. It is called a **label prefix** and is inserted automatically unless 1-2-3 is instructed otherwise. That is, it is the **default** setting.

Since 1-2-3 aligns values to the right of their cells, you may prefer column title labels to be right-aligned as well. To achieve this you type a double quote, ["], before typing your label. Position the cell pointer on **AGE:**, then enter **"AGE:** and watch the position of the label change. Notice that **"AGE:** is now displayed in the control panel.

You can alter the positions of labels already typed by using the **Range Label-Prefix** command. First position the cell pointer at one end of the range of labels to be repositioned, say on **SALARY:**. Obtain the menu, select Range and then Label-Prefix. Choose the alignment you want, which for our column titles is Right, then use the arrow keys to extend the range over the labels you want to reposition. Press [RETURN] and the realignment will take place, as shown in table 3.1. You might like to experiment with re-aligning some of the employees' names.

Table 3.1 Use of Label Prefixes

```
          A         B         C         D         E         F
1    EMPLOYEE PERSONAL DATA
2    ------------------------------
3    SURNAME:  FORENAME:  AGE:    SALARY:
4                                  (£)
5    Smith     Angela      20      4800
6    Brown     John        61     15460
7    Jones     Roberta     27      5110
8    Lee       Martin      32     10655
9    Soper     Jean        25      6325
```

You will notice that the Range Label-Prefix command offers you three label positions, Left, Right and Center. A label which is to be positioned centrally in its cell needs the prefix [^]. You can type it in as you type the label, or you can use the Range Label-Prefix command to insert it for you afterwards. Below **SALARY:** enter ^(£) as shown in table 3.1.

The Range Label-Prefix command allows you to alter the positions of labels you have already typed, but has no effect on labels you type later. You can, however, alter the label prefix which is automatically assigned by using the **Worksheet Global Label-Prefix** command. This again offers three label positions. If, using this command, you choose Right, all labels subsequently typed will be right-aligned, unless either you type a different label prefix or you reposition them with the Range Label-Prefix command. The Worksheet Global Label-Prefix command, however, has no effect on labels typed before the command is used. Remember that if labels are too wide to fit into the column in which they are entered they will overflow to the right, provided the cells to the right are empty. This is true whether the alignment given by the label prefix is Left, Right or Center.

You will see from table 3.2, which lists all the label prefix characters, that there are in fact four of them.

Table 3.2 Label Prefix Characters

```
    '  left aligned
    "  right aligned
    ^  centred
    \  repeated
```

The backslash character [\] causes the character or characters typed after it to be repeated across the width of the current cell. It is useful for drawing lines on the spreadsheet, or for drawing patterned lines comprised of 2 or 3 characters, for example, \- would appear as ---------.

If you are going to use the hyphen [-] to draw a line you must type a label prefix before typing [-], because if you simply type the hyphen, 1-2-3 will interpret it as a minus sign and decide that you are inputting a value rather than a label.

Drawing a line under the title of the worksheet you are working with will give it the form shown in table 3.1. To do this we enter a repeated hyphen in cell A2, then make the same entry in other cells in the row, using the **Copy** command. Position the cell pointer, then, on cell A2 and enter \- with the [RETURN] key. A line is drawn across the cell. Now with the pointer still positioned on the line you have drawn press the [/] key to call up the menu and **C** for the Copy command. The entry line then asks for a range to copy from, as shown in table 3.3.

Table 3.3 Control Panel of the Copy Command

```
A2: \-                                                    POINT
Enter range to copy FROM: A2..A2
```

The range suggested by 1-2-3 for copying from is the current cell. Since we do wish to copy from it, simply press the [RETURN] key. The entry line then becomes that shown in table 3.4.

Table 3.4 Control Panel after Copy From Range Selected

```
A2: \-                                                    POINT
Enter range to copy FROM: A2..A2 Enter range to copy TO: A2
```

The current cell is now suggested as the starting point of the range to copy to. We must extend this range to the last cell where we wish a line to be drawn, namely D2. Press the [.] key to anchor the range, extend it by pressing the [RIGHT] key three times, and press [RETURN]. The line is displayed as shown in table 3.1, and the mode indicator returns to READY.

Alternatively, the line could have been drawn by entering in A2 an apostrophe to denote a left-aligned label, followed by hyphens repeated across the screen as far as you wish. This method is perhaps easier; the only disadvantage is that although the line appears to extend through, say, cells B2, C2 and D2, because it is in fact entered in A2 it will be displayed only when that cell is displayed on the screen.

3.3 Overwriting and Pre-Sort

- **Data Fill**

We shall now add some further data to our spreadsheet. Enter the names and salaries shown in table 3.5.

Table 3.5 Additional Data Entries

```
         A         B          C         D         E         F
1    EMPLOYEE PERSONAL DATA
2    ----------------------------------------
3    SURNAME:  FORENAME:  AGE:   SALARY:
4                                  (£)
5    Smith     Angela      20     4800
6    Brown     John        61    15460
7    Jones     Roberta     27     5110
8    Lee       Martin      32    10655
9    Soper     Jean        25     6325
10   White     Lee               12480
11   Smith     Julia              7320
12   Lawson    Hugh               7320
13   Thatcher  Edward             8894
14   Archer    Sean               6750
15   Bingham   John              14380
16   Cox       Henry             13560
17   McKay     Iain              20860
18   Mackie    Charles            5570
19   Phillips  John               6785
20   Reed      Jane               5520
21   Young     Peter             10760
22   Baker     Rachel            14380
23   Evans     Paula              7320
```

The data on age will not be required in this chapter and so no further entries are made in that column.

We are going to sort the employee records, or the rows of our data, into ascending salary order. But **before sorting** data, **always number** it in its existing order, so that you can sort it back into its original order should you decide that you want to do so. The **Data Fill** command can be used to enter any column of regularly spaced values, so it is ideal for numbering rows. We shall **overwrite** the unwanted age data with these employee numbers and we begin by entering the new title, **Pay No:** for column C.

To use the Data Fill command, first position the pointer where the first entry is to be made, which in this case is cell C5. Call up the menu, select Data and then Fill. You will be asked to define the Fill range, which is the rows where you wish numbers to be entered. The control panel will be as shown in table 3.6, offering you the current cell as the start of the range.

Table 3.6 Data Fill Range

```
C5: 20                                              POINT
Enter Fill range: C5
```

Press [.] to anchor the range at C5 and extend it to the bottom of the existing columns of figures, then press [RETURN]. You will then be asked for the Start value, which will be entered in the first cell of the range you have just defined. The value 0 is offered, but we require 1, so type **1** and then press [RETURN].

Table 3.7 Data Fill Command

```
C5: 20                                              EDIT
Enter Fill range: C5..C23
Start: 1              Step: 1            Stop: 2047
```

Table 3.8 Pay No. Entered using Data Fill

	A	B	C	D	E	F
1	EMPLOYEE PERSONAL DATA					
2	---------	---------	-------	--------		
3	SURNAME:	FORENAME:	PAY NO:	SALARY:		
4				(£)		
5	Smith	Angela	1	4800		
6	Brown	John	2	15460		
7	Jones	Roberta	3	5110		
8	Lee	Martin	4	10655		
9	Soper	Jean	5	6325		
10	White	Lee	6	12480		
11	Smith	Julia	7	7320		
12	Lawson	Hugh	8	7320		
13	Thatcher	Edward	9	8894		
14	Archer	Sean	10	6750		
15	Bingham	John	11	14380		
16	Cox	Henry	12	13560		
17	McKay	Iain	13	20860		
18	Mackie	Charles	14	5570		
19	Phillips	John	15	6785		
20	Reed	Jane	16	5520		
21	Young	Peter	17	10760		
22	Baker	Rachel	18	14380		
23	Evans	Paula	19	7320		

We must then define the Step, which is the difference between the successive values being entered, and here we may simply press [RETURN] to accept the value 1 which is offered.

Finally we are asked for Stop, the last value to be entered. Since, however, we have accurately defined the range to be filled we may simply accept the excessively large Stop value offered. With the control panel as shown in table 3.7, we press the [RETURN] key. The numbers 1 to 19 will be displayed in column C, as shown in table 3.8.

We are now ready to sort the data.

3.4 Sorting Data

- **Data Sort**
- **Rectangular range**

The **Data Sort** command in 1-2-3 will allow us to sort our rows of data, or employee records, into either ascending or descending order of salary size. We shall choose to sort into ascending salary order. But what happens when there is more than one employee with a particular salary? If we define only the **primary** variable by which we wish to sort, namely salary, then employees with the same salary will be listed in the order in which they appear before the sorting takes place. We can, however, define a **secondary** sort criterion, in this case surname. Employees with the same salary will then be sorted according to their surname: that is, those with the same salary will be listed in alphabetical order of surname if we ask for an ascending sort.

Obtain the menu, then, and choose **Data** followed by **Sort**. The sub-menu shown in table 3.9 will appear:

Table 3.9 Data Sort Sub-Menu

```
C5: 1                                                           MENU
Data-Range  Primary-Key Secondary-Key Reset Go Quit
Specify records to be sorted
```

You must define both the Data Range and the Primary-Key; the Secondary-Key is optional.

To sort our employees into ascending salary order we must define a **rectangular Data-Range** to cover all our employee records. We choose Data-Range, move the pointer to one corner of our data, say D5, press [.] to anchor the range, then use [LEFT] and [DOWN] until all the records are covered. The range then is D5..A23, which is equivalent to A5..D23. Pressing [RETURN] enters this range as containing the records to be sorted and returns us to the sub-menu.

We now select **Primary-Key**, which invites us to:

> Enter Primary sort key address:

and offers the current cell. We can give the address of any cell in the column by which we wish to sort. In this case we might give **D5**, since column D contains the salary data. We are then asked to:

> Enter Sort order (A or D):

so type **A** for ascending, then press [RETURN].

We can then select **Secondary-Key** and give the address of any cell in the surname column to sort people with the same salary by surname. When sorting labels, an ascending sort will arrange them in alphabetical order, so you should again choose Sort order **A**.

Table 3.10 Employees Ranked by Salary

	A	B	C	D	E	F
1	EMPLOYEE PERSONAL DATA					
2	---------	-----------	---------	----------	-------	
3	SURNAME:	FORENAME:	PAY NO:	SALARY:	RANK:	
4				(£)		
5	Smith	Angela	1	4800	1	
6	Jones	Roberta	3	5110	2	
7	Reed	Jane	16	5520	3	
8	Mackie	Charles	14	5570	4	
9	Soper	Jean	5	6325	5	
10	Archer	Sean	10	6750	6	
11	Phillips	John	15	6785	7	
12	Evans	Paula	19	7320	8	
13	Lawson	Hugh	8	7320	9	
14	Smith	Julia	7	7320	10	
15	Thatcher	Edward	9	8894	11	
16	Lee	Martin	4	10655	12	
17	Young	Peter	17	10760	13	
18	White	Lee	6	12480	14	
19	Cox	Henry	12	13560	15	
20	Baker	Rachel	18	14380	16	
21	Bingham	John	11	14380	17	
22	Brown	John	2	15460	18	
23	McKay	Iain	13	20860	19	

Now that you have set the sorting instructions, select **Go** to activate the sort. The sorted data will appear as shown in the first four columns of table 3.10. Notice that the salaries are in ascending order and that when two or three employees have the same salary (£14380, £7320) their names are listed in alphabetical order.

To enable us to identify the median and quartiles we assign ranks to the ordered data. You can again use Data Fill to have the ranks inserted in column E. When you select the command the computer will respond

Enter Fill range X : C5..C23

The range offered will no doubt surprise you. It is the last range that you filled in, and 1-2-3 assumes that you may wish to fill it again. Since you do not, press the [BACKSPACE] key. You will now be offered the current cell as the range starting point, so you can move it, anchor it when you wish by pressing [.], and extend the range as desired. You should choose to fill the range E5..E23. The Start, Step and Stop values you are offered will also be those you last used, and this should be helpful to you.

Remember to use the appropriate label prefix to right align your column title, and Copy a cell containing \- to extend your line. You should then have the spreadsheet shown in table 3.10.

3.5 Minimum, Maximum, Median, Q1 and Q3

- **Doing arithmetic**
- **No spaces in formulae**
- **Cell addresses**
- **Pointing**
- **Range Name**

If we have **n** data values, once they are ranked in order of size it is easy to identify the minimum and maximum. They are the 1st and n^{th} values respectively, so in this case the minimum value is £4800 and the maximum, the 19th value, is £20860. You may like to identify them as such by entering labels in column F, as shown in table 3.11.

Since **n** is 19, (**n**+1)=20 and (**n**+1)/4= 5. The first quartile, **Q1**, is therefore the salary which is ranked 5th in order of size, and we can identify it as £6325. Similarly we can identify the median as the $(n+1)/2^{th}$ or 10th salary, namely £7320, and the third quartile, **Q3**, which is in the $3(n+1)/4^{th}$ or 15th position, as £13560.

Although mental arithmetic will suffice for finding the positions of these summary measures, we shall require to use the calculating power of the spreadsheet, and so it is useful for us to see how we can instruct 1-2-3 to do arithmetic.

Table 3.11 Minimum, Maximum, Median and Quartiles

```
           D           E         F         G           H          I
 1       ┌─────────────────────
 2       │─────────────────────
 3       │ SALARY:    RANK:                                  POSITION:
 4       │   (£)
 5       │   4800      1 Min    Min at 1                          1
 6       │   5110      2
 7       │   5520      3        Q1 at (n+1)/4                     5
 8       │   5570      4
 9       │   6325      5 Q1     Md at (n+1)/2                    10
10       │   6750      6
11       │   6785      7        Q3 at 3*(n+1)/4                  15
12       │   7320      8
13       │   7320      9        Max at n                         19
14       │   7320     10 Md
15       │   8894     11
16       │  10655     12
17       │  10760     13
18       │  12480     14
19       │  13560     15 Q3
20       │  14380     16
21       │  14380     17
22       │  15460     18
23       │  20860     19 Max
```

We can enter an **arithmetic formula** in 1-2-3 in very much the same way as we would write it by hand. Notice that the first step in entering a formula must always be to **choose a location for the result** and move the cell pointer to it. The **first character** we then type must be recognised by 1-2-3 as a VALUE. When the formula is completed and entered, 1-2-3 immediately calculates the resulting value and displays it in the highlighted cell.

Move your cell pointer to an empty cell well clear of the table you are creating, for example, in column J, and enter

    19+1

Be careful not to type any spaces between the numbers and symbols in your formula. The resulting value of of 20 will be displayed. Try entering 19-1 in another cell to display the value 18.

The first quartile position is (**n**+1)/4. Move to another empty cell and enter

    (19+1)/4

You should obtain the value 5. Notice that parentheses are required in this formula so that the addition will be done before the division. 1-2-3 has **full algebraic logic**; some other spreadsheets do not. In 1-2-3, multiplication and division are done before addition and subtraction unless you indicate otherwise by using parentheses. Notice also that 1-2-3 recognises a left parenthesis [(] as beginning a VALUE rather than a LABEL.

In 1-2-3 the slash [/] is used as the division sign, the asterisk [*] as the multiplication sign, and the hyphen [-] as the subtraction or minus sign. It is not possible to omit the multiplication sign, so that although we might write 3(**n**+1)/4 as the position of the 3rd quartile, in 1-2-3 you should enter

    3*(19+1)/4

to obtain the value 15.

We have been entering the value 19 for **n** in our formulae. But one of the important advantages of calculating with 1-2-3 stems from the fact that if a value is already displayed in the spreadsheet it can be used in a formula by referring to it by its **cell address**. The advantage of this is that should you change the value, all values calculated from it will be recalculated automatically.

The spreadsheet shown in table 3.11 has the positions in the distribution of the measures we have been calculating displayed in column I. Labels describing the formulae used to calculate the positions have been entered in column G, and spill over into column H. Enter these labels in your spreadsheet, with a title for column I. Enter the value 1 in I5 and the value of **n**, 19, in I13.

To calculate the value of (**n**+1)/4, then, which has been done in I7, we may put in our formula the cell address for **n**, I13, rather than the value itself. That is, we may enter in I7

    (I13+1)/4

to obtain the value 5.

Rather than typing in the cell address you may use the **pointer** to highlight the value in the spreadsheet which you want to use, and 1-2-3 will insert the cell address for you. Try entering in I9 your formula for the median position, (**n**+1)/2, by typing the left parenthesis then moving the pointer until I13 is highlighted. Notice as you do this that the mode indicator in the control panel changes to POINT, and that the formula successively displays the addresses of the various cells that are highlighted as you move to the desired position. When you are pointing at the value to be used, press the next key required by the formula which in this case is [+]. The pointer jumps back to the cell where the result will be entered, and the message changes back to VALUE. Finish typing your formula and press [RETURN], and the value 10 will be displayed.

There is one further way of referring in a formula to a value which is already displayed, and that is by **name**. You must first use the Range Name Create command to name the cell or cells to which you wish to refer. You may use upper or lower case letters for range names, but notice that 1-2-3 does not distinguish between them, and displays them as capitals, so you cannot give one range the name **x** and another the name **X**. Range names may also contain numbers and symbols, but it is good practice to keep them as similar to the statistical notation as possible and not to assign names which might be confused with cell addresses or with arithmetic instructions. Let us give cell I13 the name **n**. Position the pointer on I13, obtain the menu and select Range, followed by Name and then Create. You will be asked to

    Enter name:

so type **n** and press [RETURN]. The message

    Enter range: I13..I13

is then displayed, offering you the current cell as the range to which this name is to be applied. Since you wish to accept this, press [RETURN]. I13 is now named **n**, and may be referred to as such in formulae.

Let us enter in I11 a formula to calculate the position of the third quartile. Move the pointer to I11 then type

    3*(n+1)/4

On entry, the value 15 will be displayed. Range names allow us to enter formulae in 1-2-3 very much as we would write out the definition of what we wish to calculate. They therefore make 1-2-3 formulae much more readable, but each name can only be used to refer to a single cell except with built in functions which we shall meet in chapter 4.

You should now have completed the spreadsheet shown in table 3.11. Remember to save it under a new filename so it is available for future use.

3.6 Summary

In this chapter we have seen how to improve the layout of a spreadsheet by choosing appropriate label prefixes to position our labels. We can use the Worksheet command to set the default prefix and we can alter prefixes already assigned using the Range command. Notice that we cannot align values in this way. Numbers preceded by label prefixes are treated as labels and cannot be used in calculations.

We have entered a sequence of numbers using Data Fill, and discovered how easy it is to reorder a set of data in the way we specify by using Data Sort. This enabled us to identify various measures of position for our data, namely the minimum, maximum, median and quartile values. We have used 1-2-3 formulae to do simple arithmetic and have noted several things about these:

1-2-3 Formulae:

- Must begin with a character recognised by 1-2-3 as a VALUE prefix.

- Must not contain any spaces.

- Should refer to values already entered by their cell addresses.

- Can have the cell addresses they contain selected by pointing.

- Can use a range name rather than a single cell address to make them more readable.

4
Mean and Standard Deviation

- Editing mistakes
- Worksheet Erase
- Formatting values
- Adjusting column widths
- Formulae : using a range of displayed values
- Calculations in stages
- Copying formulae
- "What if?" calculations

4.1 Introduction

In the last chapter we identified various measures of position by which a data set can be summarised. Alternative summary measures are provided by the **mean**, or average, which may be regarded as typifying the data, and by the **standard deviation**, which is a measure of the spread of the data values.

In this chapter we shall use Lotus 1-2-3 to find these measures for a set of data values. We shall break the calculations down into manageable chunks, and input formulae to perform the computations. To carry out the same calculation on a column or row of values we shall use the Copy command. For some calculations 1-2-3 offers built in functions which we shall compare with our basic methods.

We have already seen how to align labels by using label prefixes. This chapter begins by showing you how to alter the alignment of values, which is done by selecting appropriate menu commands so as to specify the

format in which the values are to be displayed. If a value cannot be displayed in the specified format in the existing column width, a row of asterisks is shown. Widening the column will enable the number to be displayed. We shall look, therefore, at the commands which adjust **column width**, and shall find that they are generally useful for improving the appearance of the spreadsheet.

We shall be starting a new worksheet in this chapter. If your last spreadsheet is still displayed, first check that you have saved it, then use the Worksheet Erase command to clear the screen. To do this, press [/] to obtain the menu then select Worksheet Erase Yes. The displayed worksheet will disappear, and you would only be able to get it back by retrieving the file you have previously saved.

4.2 Formatting Values

- Function key [2/EDIT]
- Worksheet Global Format
- Range Format

So far, the values we have entered in 1-2-3 have all been integers (whole numbers) and have been displayed as such. They have been aligned to the right of their cells so that the unit figures for the various values have been directly below one another. The default format setting, General, has been appropriate for our data. When this is not so, however, as in the following example, we can choose the appropriate format with the Worksheet or Range commands.

Begin a new worksheet and enter the data on World Bank Lending, placing the source of information in row 19 to allow working space, as shown in table 4.1. If you notice a mistake in something you have already entered in the spreadsheet, you can enter EDIT mode to alter it. First position the pointer on the cell you wish to change, then press function key **[2/EDIT]**. The message indicator will change to EDIT, and the contents of the cell, followed by the cursor, will be displayed in the second line of the control panel. You may move the cursor with the [LEFT] and [RIGHT] arrow keys; the [BACKSPACE] key will delete the character preceding the cursor; and any characters typed will be inserted at the cursor position. When you have finished, press [RETURN] to insert the revised entry in the spreadsheet. Note that if you try to enter something that 1-2-3 does not understand, for example, **1980s**, which it thinks is a VALUE, but then finds contains a letter, it will place you automatically in EDIT mode.

Table 4.1 Loan Entries

```
         A       B          C         D          E
1     World Bank Lending to Borrowers in Western Africa
2     for Water Supply and Sewage     (US $ million)
3
4     Year       Amount, X
5
6         1980         21
7         1981          5
8         1982        110
9         1983       31.5
1         1984       10.9
11
12
13
14
15
16
17
18
19    Source: World Bank Annual Report 1984
```

We notice that with the default format, **General**, only as many decimal places as are required are shown. Consequently, as we can see in table 4.1, the values in a column may not be correctly aligned with the unit figures lying underneath each other. To alter this we must specify the format in which the values are to be displayed.

In 1-2-3 we can set the display format for values, either for the whole worksheet using **Worksheet Global Format**, or for any range of cells which we specify, using **Range Format**. When a Format command is used, the format specification is attached to the relevant cells. Any values already in these cells and also any values subsequently entered there are displayed in accordance with it. When a cell which has been formatted is active, the format chosen is shown in parentheses in the first line of the control panel, preceding the contents of the cell, if any.

The numerical value stored by the computer is not altered by the format chosen to display it. Nor are labels entered in formatted cells affected in any way.

Let us use the Range Format command to reformat the data values in column B. At present we have values in rows 6 to 10, but if we format rows 1 to 19 we shall find this has no effect on our labels, and that other values which we are going to calculate in the column will be displayed in the chosen format.

Remember that when you are going to define a range it is advantageous to move the pointer to an end point of that range before beginning the command. Move the pointer, then, to B1, and call up the menu. Now you are becoming familiar with 1-2-3, you will probably find it quickest to select commands from it by typing their initial letters. Type, then, **R** for Range followed by **F** for Format. The screen control panel will then show:

Table 4.2 The Format Sub-Menu

```
B1:                                                           MENU
Fixed  Science Currency , Gen. +/- Percent Date Text Reset
Fixed number of decimal places (x.xx)
```

The format sub-menu, shown in table 4.2, lists the various display formats which 1-2-3 offers. As we have already noted, the default setting is the **General** format, which suppresses trailing zeros after the decimal point. It also expresses very large or very small numbers in Scientific format. **Reset** allows you to return to this format having previously chosen another.

The **+/-** format allows you to display a horizontal bar graph. **Text** format shows formulae as they were typed rather than displaying their numerical result, and **Date** is for dates. The remaining formats offered in the sub-menu require the user to set a fixed number of decimal places between 0 and 15, with the default setting being 2. **Fixed** format, then, simply displays the value with the specified number of decimal places. The [,] format in addition places commas between three-digit groups, and encloses negative values in parentheses. The **Currency** format is the [,] format preceded by a [$] sign. **Percent** shows the value divided by 100 and followed by [%]. **Scientific** notation allows a very large or small value to be displayed by expressing it as a number to be multiplied by 10 to the power of the exponent shown. For example, 4.7E-01 is 4.7 multiplied by 10^{-1}, which is 0.47.

We select Fixed format by pressing **F**, and 1-2-3 responds with the control panel shown in table 4.3.

Table 4.3 Format Fixed Command: No. of Decimal Places

```
B1:                                                           EDIT
Enter number of decimal places (0..15): 2
```

Since 2 places are offered we simply press [RETURN] and the control panel becomes that shown in table 4.4.

Table 4.4 The Format Fixed Command - Range

```
B1:                                                      POINT
Enter range to format: B1..B1
```

We want to reformat the range B1 to B19, so extend the offered range by moving the pointer to B19 and pressing [RETURN]. The screen is then as shown in table 4.5, with the Amount values now aligned.

Table 4.5 Amount Values Reformatted

```
B1: (F2)                                                 READY
```

```
         A         B          C         D         E         F
1   World Bank Lending to Borrowers in Western Africa
2   for Water Supply and Sewage    (US $ million)
3
4    Year      Amount, X
5
6         1980      21.00
7         1981       5.00
8         1982     110.00
9         1983      31.50
10        1984      10.90
```

Notice that the format you have set for the active cell is shown in parentheses in the top line of the control panel. It comes between the cell reference and the cell contents, if any. In this instance the label in cell A1 spills over into B1 which is actually blank. The format message (F2) indicates that Fixed format with 2 decimal places has been chosen.

Note that if we had wanted to reformat the whole spreadsheet the same set of format options except Reset are available from the Worksheet Global Format command. This command reformats all cells, with the exception of any whose formats have already been specified using Range Format.

4.3 Adjusting Column Width

- **Worksheet Global Column-Width**
- **Worksheet Column-Width**

Associated with aligning labels and setting the format of values is selecting the display **column width**. The default width setting is 9 characters per column. For values this allows 8 digits to be displayed, with the rightmost character being blank as a separator. The screen can then show 8 columns simultaneously. If the data and titles will fit into narrower columns, setting smaller widths will enable more columns to be simultaneously displayed. Numbers which, in their specified format, are too large to fit in the existing column width are indicated by asterisks. They can be displayed by widening the column.

The Column-Width command is available under Worksheet Global for resetting all columns in the worksheet, and is also itself in the Worksheet sub-menu for setting the width of individual columns. Let us widen column B by one character space so that its label does not extend into column C. Press [/] for the menu, W for Worksheet and C for Column Width. The screen will display what is shown in table 4.6.

Table 4.6 The Worksheet Column-Width Sub-Menu

```
B1:  (F2)                                              MENU
 Set   Reset
Set width of current column
```

Press [RETURN] to accept the command Set which is offered, and the control panel changes to that in table 4.7.

Table 4.7 The Worksheet Column-Width Set Command

```
B1:  (F2)                                             POINT
Enter column width (1..72): 9
```

Notice that the mode indicator has altered to POINT, indicating that the column can be widened by the [RIGHT] arrow key and narrowed by the [LEFT] one, with the reformatted column being immediately displayed. Press the [RIGHT] arrow key once and the value in the entry line changes to 10. Now press the [RETURN] key to complete the setting procedure. Instead of using the pointer, the number of characters required in the column could have been typed in directly, but with this method the reformatted column is not displayed until the [RETURN] key is pressed.

We are now ready to begin the statistical calculations.

4.4 The Arithmetic Mean

- **Using displayed values in a formula**
- **Beginning formulae with [+]**
- **[@] functions**
- **Naming a range of values.**

The **arithmetic mean**, or simply the **mean**, or **average** of a set of values, **X**, is defined as the sum of those values divided by the number, **n**, of values in the set. If we denote it **Xbar**, we may write:

$$\text{Xbar} = \sum X / n \qquad [4.1]$$

To find the mean amount lent per year by the World Bank to borrowers in Western Africa over the period 1980 to 1984, we must add the amounts in column B and divide their total by the number of years for which there are data, namely 5.

We shall form the total in cell B12, so first move the pointer to the adjacent cell, A12, and enter an appropriate label, **Total =**. If you finish this using the [RIGHT] arrow key, the pointer will move to B12, which is the cell where you wish to enter a formula to form the total of the amounts, **X**.

Remember that when we include displayed values in formulae we want to use either their **cell addresses** or **names** which we have given them. There are several reasons for doing this instead of retyping the values into the formulae. Firstly, it avoids possible typing mistakes. Secondly, it ensures that any decimal places being stored by the computer but not displayed are in fact used. Next, it allows the formula to be copied to other cells in which a similar calculation is to be made using values in corresponding positions in the spreadsheet. And lastly it allows sensitivity analysis to take place since all values calculated from a particular value will automatically be recalculated if it is altered.

To total the values in column B, then, the basic formula will contain the five cell addresses each connected to the next by a [+] sign. But if we try to begin our formula by typing, say, **B6**, 1-2-3 will assume that the letter **B** indicates a label rather than the value which we want. To prevent this we must preface the cell address by a [+] sign.

Beginning an entry with a [+] indicates to 1-2-3 that you are entering a VALUE, and you may then obtain a cell address by pointing. It is good practice to form the habit of always beginning formulae with a [+] (or a [-] if appropriate), except when they start with [@], which is the symbol that begins an inbuilt function in 1-2-3.

To form the column total in B12, then, enter a [+] to tell 1-2-3 that you are forming a value. Next, use the [UP] arrow to move to the first of the values to be included in the sum. The order in which the cell addresses appear in the formula is, of course, of no importance. When, say, the amount 10.90 in cell B10 is highlighted, press the [+] key to inform 1-2-3 that another value is to be added to it. The pointer returns to the cell in which the formula is being entered. You may then repeat the procedure of moving it to the next amount to be included in the sum and pressing [+] if another value is to be added. When the last value is highlighted the formula is complete and should appear as: +B10+B9+B8+B7+B6. Press the [RETURN] key to compute the result and enter it in the table. The value 178.40 should be displayed, as shown in table 4.8.

To form the mean we move to another cell and divide the total we have formed by 5. Notice that we could have written a formula to find the mean directly, but it is one of the features of a spreadsheet that you may perform calculations one step at a time so that you can see directly how each stage in the computation takes place. We shall place the mean in B13, below the total value, and so first enter an explanatory label, **Total/5 =**, in A13. We then enter the formula

    +B12/5

in B13, obtaining the cell address by pointing, using [/] as a division sign and following it with the number 5. This completes the calculation of the mean using the definitional formula, and gives the worksheet shown below.

Table 4.8 Total obtained by Addition Formula

```
              A         B         C         D         E         F
 1   World Bank Lending to Borrowers in Western Africa
 2   for Water Supply and Sewage     (US $ million)
 3
 4   Year        Amount, X
 5
 6         1980       21.00
 7         1981        5.00
 8         1982      110.00
 9         1983       31.50
10         1984       10.90
11
12   Total =         178.40
13   Total/5 =        35.68
```

Once you understand the computational procedure you may wish to simplify your formula writing by utilising the **inbuilt functions** offered by 1-2-3. Their use is signalled by the [@] key, which is followed by letters defining the function and then by parentheses enclosing certain information required for the calculation of the function. This information consists of one or more values or ranges of values. These are called the **arguments** of the function, and they are separated by commas when more than one are required. Arguments are usually defined by their cell addresses or by the range names which they have been given. We shall now use @SUM and @AVG to find respectively the total and mean of a range of values.

You may like to begin by creating a name for the range of values for which we want to calculate these measures, namely the amounts, **X**. As usual when you are going to use a command in which a range is defined you will find it easiest to move the pointer beforehand to the starting point of the range, say to B6. After calling up the menu, select **Range Name Create** and type the name you wish to give the range, say **X**, in response to the computer's prompt. When you press [RETURN] the message

Enter range: B6..B6

will appear. Extend the range to cover all the amount values and press [RETURN]. The chosen name will be assigned to the designated range of cells.

So that we can compare the values computed by the [@] functions with those we have already calculated, we move to a fresh part of the spreadsheet and enter an identifying label, **SUM =**, in cell A15. Notice that this label begins with a letter. If you wish to begin with [@] remember to type a label prefix or 1-2-3 will assume you are beginning a VALUE.

Locate the pointer in B15 and enter the formula

@SUM(B10..B6)

where the range enclosed in parentheses can be obtained by pointing. Notice that less movement of the pointer is required when the range is defined from bottom to top. Alternatively, if the argument range has been named **X** you may type your formula as

@SUM(X)

In either case, when the [RETURN] key is pressed, 1-2-3 calculates and displays the result, which we note from table 4.9 is identical with our previous total.

Table 4.9 @SUM and @AVG Functions Used

```
           A         B           C        D        E        F
 1   World Bank Lending to Borrowers in Western Africa
 2   for Water Supply and Sewage    (US $ million)
 3
 4   Year         Amount, X
 5
 6        1980         21.00
 7        1981          5.00
 8        1982        110.00
 9        1983         31.50
10        1984         10.90
11
12   Total =           178.40
13   Total/5 =          35.68
14
15   SUM =             178.40
16   AVG =              35.68
17
18
19   Source: World Bank Annual Report 1984
```

The function @AVG allows the mean to be calculated directly. Enter a label, **AVG =**, in A16 and move the pointer to B16. Enter the formula

@AVG(B10..B6) or @AVG(**X**)

and again check that the result obtained corresponds with our previous value for the mean.

4.5 The Standard Deviation

- Calculations in stages
- Relative address
- Absolute address
- Copying formulae
- Exponentiation
- Function key [4/ABS]

The **standard deviation** measures the spread of values about the arithmetic mean. For a particular set or population of values, it is defined as:

$$\text{Standard deviation} = \sum (\mathbf{X} - \mathbf{Xbar})^2 / \mathbf{n} \qquad [4.2]$$

(see Hodge and Seed, p. 73), and in a spreadsheet we can work directly with this definitional formula. Because of the steps involved in its calculation it is sometimes called the **root mean squared deviation**. When the values from which it is calculated form a sample and the result is to be used as an estimate for the population standard deviation, the divisor **n - 1** is substituted for **n** in the above formula (Hodge and Seed, p. 158).

We shall carry out the calculation in stages starting at the innermost level of parentheses. The first step is to form in column C a column of values of deviations from the mean. Move the pointer to C4 and enter a suitable title label. Each of the values in column C is to be the value in the corresponding row of column B, minus the mean, for which you can use either value displayed. Let us use the value in B13. If we enter an appropriate formula in C6 we shall be able to use the Copy command to calculate the whole column of values.

Unfortunately, there is no way of instructing 1-2-3 to perform the same arithmetic operation on each of the values in a named range. You may like to name the mean **Xbar** and use it instead of the cell address in your arithmetic instructions. The various amounts in the range named **X**, however, must each be referred to by their separate cell addresses.

Begin by moving the pointer to C6, then type [+] to indicate to 1-2-3 that you are entering a VALUE. Now point at the first amount, **X**, in B6. Your formula will read

    +B6

If you enter this formula and use the Copy command to copy it down the column, as we did in the last chapter for labels, you will find each row of column C then contains the same value as the corresponding row in column B. That is, as the address B6 has been copied from one row to the next, the row number has changed accordingly each time. The address B6 is a **relative address**; the row and column are considered relative to the cell in which the formula is being entered.

Cell addresses are assumed by 1-2-3 to be relative, unless you instruct it otherwise. When we subtract the mean from our column of amount values we wish to refer always to one specific cell, B13, containing the mean. We can do this by giving it an **absolute** address, which is indicated by [$] signs preceding both the column letter and row number, or by a [$] sign preceding the range name, **Xbar**.

If you have entered your part-finished formula

    +B6

press the [2/EDIT] function key to allow you to add to what you have typed, otherwise simply continue typing. Your completed formula should read:

+B6-$B$13

If you wish, you may type the keys required. Pointing to obtain the cell address, however, enables you to highlight the value you are going to use in your formula. You should do this as you have done before: simply type the [-] sign, then point at B13. If you then, before typing anything else, press the **[4/ABS]** function key once, the address will be made absolute by the insertion of [$] signs. If you were to press the [4/ABS] key again, you would obtain an address which was part absolute, part relative, as you would if you pressed it a third time. Pressing it a fourth time brings you back to the relative address you started with, and continuing to press it would repeat the cycle.

Alternatively, if you have named cell B13 as **Xbar**, you may complete your formula to read

+B6-$**Xbar**

Notice that you will have to type the [$] sign followed by the name you have given the cell to ensure an absolute cell address. Now enter your formula and replicate it down to C1O using the Copy command. You should obtain the column C values shown in table 4.10.

Table 4.10 Standard Deviation Calculated

```
         A         B          C        D         E         F
 1   World Bank Lending to Borrowers in Western Africa
 2   for Water Supply and Sewage    (US $ million)
 3
 4   Year       Amount, X    X-Xbar  Sq(X-Xbar)
 5
 6        1980      21.00   -14.68   215.5024
 7        1981       5.00   -30.68   941.2624
 8        1982     110.00    74.32  5523.462
 9        1983      31.50    -4.18    17.4724
10        1984      10.90   -24.78   614.0484
11
12   Total =       178.40     .00   7311.75
13   Total/5 =      35.68           1462.35
14
15   SUM =         178.40           Root Mean
16   AVG =          35.68           Squared Deviation
17                                    38.24067
18
19   Source: World Bank Annual Report 1984
```

The next step is to square each of the deviations we have just formed. We shall place the squared values in column D, and label D4 accordingly. The formula we enter in D6 uses a relative address to refer to the deviation in the same row of column C. It is then suitable for copying. We may use a formula which multiplies the value by itself:

+C6*C6

or we may exponentiate to the power of 2:

+C6^2

Notice that the symbol [^] indicates **exponentiation**. The number to its left is raised to the power of the number to its right. Copying either formula from D6 down to cell D10 yields the values shown in the column in table 4.10. Since we have not formatted columns C and D the values are shown in General format. Note that this format allows the number of decimal places displayed to be contracted as necessary for the column width set, so that 5523.462 is shown with only 3 decimal places.

We require to sum the values we have placed in column D and to divide by the number of values, **n**, that there are. That is, we have to find the mean of the column D values. We already have, in cell B13, a formula for the mean of a corresponding column of values and we may copy it across to D13. For completeness, we copy also the formula for a column total from B12 to C12 and D12. The formulae we are copying are in cells formatted to display 2 decimal places, and this format is automatically copied across also. We obtain the values shown in table 4.10. Note that the sum of the deviations about the mean in column C is zero. It is a property of the mean that this should be so (Hodge and Seed, p72). The mean of the squared deviations shown in cell D13 is called the **variance** of the **X** values.

The final step in our calculation is to find and place in D17 the square root of the variance which we have formed in cell D13. For this we use one of the inbuilt mathematical functions, @SQRT. It requires the address of the value whose square root is to be found to be given in parentheses, immediately following the statement of the function. The formula is then:

@SQRT(D13)

which computes and displays the value of 38.24067 shown in table 4.10, which is the standard deviation we require. The identifying label entered in cells D15 and D16 describes the method of calculation.

There is, in fact, an inbuilt function which computes the standard deviation by the formula we have used. Notice that the divisor used is **n**;

in some other computer packages **n**-1 is used as the divisor. The function is defined as @STD, with the range of values whose standard deviation is to be calculated given in parentheses. Enter a label in A17, and the formula:

@STD(B10..B6)

in B17. Remember that after typing the left parenthesis, [(], you can point to indicate the range. Notice that B17 has been formatted, so only 2 decimal places are displayed. Table 4.11 shows the resulting spreadsheet after columns C and D have been widened, and D has been formatted to show 3 decimal places.

Table 4.11 Using the @STD Function

	A	B	C	D	E	F
1	World Bank Lending to Borrowers in Western Africa
2	for Water Supply and Sewage (US $ million)
3						
4	Year	Amount, X	X-Xbar	Sq(X-Xbar)		
5						
6	1980	21.00	-14.68	215.502		
7	1981	5.00	-30.68	941.262		
8	1982	110.00	74.32	5523.462		
9	1983	31.50	-4.18	17.472		
10	1984	10.90	-24.78	614.048		
11						
12	Total =	178.40	.00	7311.748		
13	Total/5 =	35.68		1462.350		
14						
15	SUM =	178.40		Root Mean		
16	AVG =	35.68		Squared Deviation		
17	STD =	38.24		38.241		
18						
19	Source: World Bank Annual Report 1984

When you have completed your spreadsheet remember to save it under a suitable file name so that it is available for future use.

4.6 "What If?" Calculations

The 1982 value of $110 million is much larger than the values for any of the other years, and we may be interested to see how different the mean and standard deviation would have been if this exceptional value had not occurred. It is one of the important features of a spreadsheet that we may **change** any value displayed, and all values calculated from it will be immediately **recalculated** also. Before beginning a "What if?" calculation be sure you have saved a copy of your original spreadsheet.

Let us suppose that the amount lent in 1982, in millions of dollars, had been 55 rather than 110. We move the pointer to B8 and enter the value 55, which replaces the previous value. Immediately the various calculations are re-computed using this new value, and the spreadsheet shown in table 4.12 is displayed. We notice that the mean is reduced to about two-thirds of its previous value, and that the standard deviation is now less than half what it was before.

Table 4.12 "What If?" 1982 Value Altered

```
         A           B          C          D          E          F
 1  World Bank Lending to Borrowers in Western Africa
 2  for Water Supply and Sewage    (US $ million)
 3
 4  Year       Amount, X    X-Xbar     Sq(X-Xbar)
 5
 6      1980      21.00     -14.68        215.502
 7      1981       5.00     -30.68        941.262
 8      1982      55.00      30.32        919.302
 9      1983      31.50      -4.18         17.472
10      1984      10.90     -24.78        614.048
11
12  Total  =     123.40        .00       1556.548
13  Total/5 =     24.68                    311.310
14
15  SUM =        123.40            Root Mean
16  AVG =         24.68            Squared Deviation
17  STD =         17.64                  17.644
18
19  Source: World Bank Annual Report 1984
```

"What if?" calculations are useful for letting you get the "feel" of a set of data and for understanding what it is that statistical calculations made from the data really measure.

4.7 Summary

In this chapter we have seen how to format numerical values and adjust the width of columns. Together with the use of label prefixes these commands give us considerable scope for altering the appearance of our spreadsheet. We may adjust particular areas of the spreadsheet using Range Format and Worksheet Column-Width. More generally, the display of all cells, except any already set by the two commands just described, can be altered by using Worksheet Global Format and

Worksheet Global Column-Width. These commands alter the display of items already entered, and they will also affect the display of subsequent entries.

As we have made further use of formulae for calculating values we have begun them in different ways. For reference, the complete set of characters which 1-2-3 recognises as beginning a value is given in table 4.13. Other characters are assumed to begin a label, except for the slash, [/], which calls up the menu.

Table 4.13   Characters Recognised as Beginning a Value

0 1 2 3 4 5 6 7 8 9 . + - (@ # $

Notice that in addition to those characters we have already mentioned, the list in table 4.13 includes the decimal point [.], the [$] to indicate an absolute address and the hash [#] which is a logical operator and which we have not yet used.

Sometimes our formulae have involved a range of values. When an [@] function is available for the calculation, an argument range can be given as a range name. When, however, we want to perform a particular arithmetic operation on each of the values in a range we cannot specify this using the range name, but must use the individual cell address.

We have seen that formulae can be copied to other cells of the spreadsheet and that as this takes place relative addresses will be altered accordingly, while those that have been made absolute by the insertion of [$] signs will remain unchanged. Other features of 1-2-3 are also becoming apparent. We have seen that calculations can be laid out in stages, and recalculation can take place to answer "What if?" type questions.

5
Frequency Distributions

- Worksheet Delete
- Range Erase
- Data Distribution
- Worksheet Insert

5.1 Introduction

A **frequency distribution** lists the number of items of data that fall into each of various numerical categories. In this chapter we shall use the Data Distribution command to form a frequency distribution for one variable from a set of observations or data records. The distribution is sometimes more meaningful if the frequencies are expressed as proportions or percentages of the total frequency, thus forming a **relative frequency distribution**. It is easy to construct this with 1-2-3. We shall also form **cumulative frequency distributions** which show the numbers or percentages of items which are less than the upper limit of each category. All these distributions can be depicted pictorially using 1-2-3's Graph commands.

The observations for which we shall construct a frequency distribution are the salary values in the Employee Personal Data spreadsheet which we used in chapter 3. If you retrieve your final version of that spreadsheet part of which is shown in table 3.11 you will find it identifies various summary measures which are not necessary for our frequency distribution. We have already saved a spreadsheet containing them, so you may like to remove them from your present worksheet before beginning any new calculations. We shall see how to do this using either the **Worksheet Delete** or **Range Erase** commands.

5.2 Blanking Spreadsheet Cells

- **Range Erase**
- **Worksheet Delete**
- **Save before using these commands**

The more you find you can get a spreadsheet looking the way you want it to, the more confident you will feel in using it. We have already found that it is easy to replace the entry in a cell with a new one. There is no need to erase a previous entry before making another one.

If you enter a space in a cell, 1-2-3 interprets it as a label and the cell appears blank. This can be a quick way of removing entries and giving a spreadsheet the appearance you want. The cell, however, is not empty, although it appears so, and this can cause problems when using certain of the data commands.

To blank a range of cells which have been filled you should use one of the menu commands. **Range** from the main menu followed by **Erase** is available for blanking any range of cells you specify; if the range to be left blank forms a complete column or row it is perhaps easier to use **Worksheet**, selecting **Delete** then Column or Row as appropriate. Remember that if you are not sure which command to select from the menu you can move along it with the [RIGHT] arrow key, displaying each sub-menu in turn. Once you have selected a sub-menu you can similarly preview each option available from it. If you choose wrongly, you can go back to the previous menu level by pressing the [ESC] key.

Having retrieved our last version of the Employee Personal Data Spreadsheet, partly shown in table 3.11, let us start by deleting the pay numbers in column C. We would like to remove this column completely, moving all entries on its right one column to the left. To do this we use the Worksheet Delete Column command. As usual, we shall find it convenient to begin by moving the pointer, in this case to any cell in column C.

Next obtain the menu and select successively Worksheet, Delete and Column. The prompt now asks for a range to delete, offering the current cell, as shown in table 5.1. Note that an entire column or columns will be deleted. The range of columns must include row numbers, but these will be ignored.

Table 5.1 The Worksheet Delete Column Command

```
C1:                                                          POINT
Enter range of columns to delete: C1..C1
```

Since we simply wish to delete column C, just press [RETURN].

All entries in column C are removed, and those to the right are shifted to fill the empty space, so that column C now contains the salary values, as shown in table 5.2. Any addresses of cells which are used in formulae are adjusted if necessary.

Table 5.2 Old Column C Removed by Worksheet Delete

```
           C         D        E         F          G         H
 1
 2    |-----------------
 3    | SALARY:   RANK:                        POSITION:
 4    |  (£)
 5    |  4800      1 Min    Min at 1                       1
 6    |  5110      2
 7    |  5520      3        Q1 at (n+1)/4                  5
 8    |  5570      4
 9    |  6325      5 Q1     Md at (n+1)/2                 10
10    |  6750      6
11    |  6785      7        Q3 at 3*(n+1)/4               15
12    |  7320      8
13    |  7320      9        Max at n                      19
14    |  7320     10 Md
15    |  8894     11
16    | 10655     12
17    | 10760     13
18    | 12480     14
19    | 13560     15 Q3
20    | 14380     16
21    | 14380     17
22    | 15460     18
23    | 20860     19 Max
```

Let us now use Range Erase to remove all entries to the right of the new column C. Move the cell pointer to one corner of the range to be erased, say D1. Call up the menu and choose Range followed by Erase. The computer will ask for a range to erase, as shown in table 5.3.

Table 5.3 The Range Erase Command

```
D1:                                                    POINT
Enter range to erase: D1..D1
```

Extend the range to cover all items you wish to erase, and press the [RETURN] key. The cell entries are removed and the spreadsheet becomes as shown in table 5.4.

Table 5.4 Entries Removed by Range Erase

	A	B	C	D	E	F
1	EMPLOYEE PERSONAL DATA					
2	-------------------------					
3	SURNAME:	FORENAME:	SALARY:			
4			(£)			
5	Smith	Angela	4800			
6	Jones	Roberta	5110			
7	Reed	Jane	5520			
8	Mackie	Charles	5570			
9	Soper	Jean	6325			
10	Archer	Sean	6750			
11	Phillips	John	6785			
12	Evans	Paula	7320			
13	Lawson	Hugh	7320			
14	Smith	Julia	7320			
15	Thatcher	Edward	8894			
16	Lee	Martin	10655			
17	Young	Peter	10760			
18	White	Lee	12480			
19	Cox	Henry	13560			
20	Baker	Rachel	14380			
21	Bingham	John	14380			
22	Brown	John	15460			
23	McKay	Iain	20860			

This is all that happens when Range Erase is used. Other cell entries are not moved to fill the gap, nor are any adjustments made to existing formulae. In other words, only the contents of the cells are removed.

If, then, we had used Range Erase earlier to remove the column entries in column C we should have been left with a blank column, whereas using Worksheet Delete Column closed up the gap.

Notice that the Range Erase and Worksheet Delete commands are powerful and destructive. To get back to your original spreadsheet now you must retrieve it again from disc. If you make a mistake you may wish to do this. It is a useful safeguard, therefore, to **always ensure you have saved** a copy of your spreadsheet before you start using these commands.

5.3 Forming Frequency Distributions

- **Data Distribution**

A **frequency distribution** lists how many of the values in a set of observations fall into each of various numerical categories. To form one from the raw data you must decide on the classes you wish to use and count how many of the observations fall into each. The **Data Distribution** command in 1-2-3 will do the counting automatically for you, if you first provide it with a column containing the **upper limits** of each class. The frequencies, when they are calculated, will be placed alongside these limits, and one row below the last of them, in the column to the right. Appropriate space must, therefore, be available in that column.

If you choose equal class intervals you can set up the column of upper limits using the Data Fill command. Equal intervals also make the plotting of a histogram easy.

Let us, then, set up equally spaced upper class limits in column D so that the frequencies will appear in column E. Enter titles for these columns, move the pointer to D6, obtain the menu and select Data Fill as before. When the computer responds

    Enter Fill range: D5..D23

you will probably be surprised at the range it offers you. It still remembers the last range you filled with ranks when you worked on the spreadsheet in chapter 3, and it offers you this even though it is now moved from column E to column D, because of the deletion of column C. If you are happy with the starting point of this range, you can accept it. The range is larger than we require, but if we define exactly the value at which the Fill process should stop, the rest of the range will just remain unfilled. If you prefer, as shown in table 5.6, to start the class limits in D6, press [BACKSPACE] to rub out the offered range. You will then be offered the current cell, D6, as a starting point. You can pin and extend the range until it is at least as long as you require. The sizes of the salary values suggest that a starting upper limit of 5000 and a step of 2500 will give a suitable number of classes. The Data Distribution command counts the number of values which are not greater than the upper limit of a class, but which are greater than the upper limit of the previous class. For the first class, of course, there is simply no previous upper limit. To ensure that all the values are included in the distribution formed, the command automatically adds a last class into which are put any values greater than the last upper limit you define. You can therefore select a stop value for the Data Fill command which is just less than the largest value in the distribution, say 20,000.

With the control panel under the Data Fill command as in Table 5.5. you can press the [RETURN] key

Table 5.5 Data Fill Command for Upper Class Limits

```
D6:                                                              EDIT
Enter Fill range: D6..D20
Start: 5000              Step: 2500              Stop: 20000
```

This enters the values shown in column D of table 5.6.

Table 5.6 Frequency Distribution Formed

	A	B	C	D	E
1	EMPLOYEE PERSONAL DATA
2	--------------------------
3	SURNAME:	FORENAME:	SALARY:	UPPER CLASS	
4			(£)	LIMIT	f
5	Smith	Angela	4800		
6	Jones	Roberta	5110	5000	1
7	Reed	Jane	5520	7500	9
8	Mackie	Charles	5570	10000	1
9	Soper	Jean	6325	12500	3
10	Archer	Sean	6750	15000	3
11	Phillips	John	6785	17500	1
12	Evans	Paula	7320	20000	0
13	Lawson	Hugh	7320		1
14	Smith	Julia	7320		
15	Thatcher	Edward	8894		
16	Lee	Martin	10655		
17	Young	Peter	10760		
18	White	Lee	12480		
19	Cox	Henry	13560		
20	Baker	Rachel	14380		
21	Bingham	John	14380		
22	Brown	John	15460		
23	McKay	Iain	20860		

We can now again call up the menu and form the frequency distribution by selecting **Data** followed by **Distribution.** The computer requests a Values range, which is the set of values, in this case the salaries, which are to be formed into a distribution. When this has been entered, the computer asks for a Bin range. The bins are the classes, defined by their upper limits, so the Bin range should consist of the upper limits we have entered in column D. If you have set up your spreadsheet as in columns A to D of table 5.6, your control panal for the Data Distribution command should be as in table 5.7.

Table 5.7 Data Distribution Command

```
D6:  5000                                              POINT
Enter Values range: C5..C23      Enter Bin range: D6..D12
```

On pressing the [RETURN] key the frequencies, **f** appear in rows 6 to 13 of column E, as shown in table 5.6.

These values which have been entered using the Data Distribution command can now be treated just like any other values. Try entering in E15, as shown in table 5.8, a formula to find their sum. Notice that because the frequencies have been entered by a command rather than by a formula, they will not automatically recalculate if you change some of the values from which they have been formed. You will need to repeat the Data Distribution command.

The last frequency is shown without a corresponding class limit. Following the pattern of these limits it is appropriate to enter a value of 22500 in cell D13. This is shown in table 5.8.

5.4 Relative Frequencies

- **Formulae for values better than commands**
- **Copying a formula to a rectangular range**

A **relative frequency distribution** shows the class frequency, **f**, as a proportion of the total frequency, Σf. That is, the set of relative frequencies are defined as

$$\text{Relative frequency} = f/\Sigma f \tag{5.1}$$

For our first salary class which has an upper limit of 5000 the relative frequency is 1/19, as can be seen from column E of table 5.8. It may be expressed in 1-2-3 either as a percentage, 5.26%, as shown in column F, or in decimal format as 0.0526. Relative frequencies are particularly useful when two frequency distributions are to be compared.

To form a relative frequency distribution, then, in column F of our spreadsheet we first enter a title then move to F6 to calculate the first relative frequency. We shall want to copy down the column the formula which we enter. The cell address for the frequency, **f**, should therefore be relative so that it will change as the formula is copied from one row to the next. The divisor, however, must always be the same value, Σf, so its address must be absolute.

A suitable 1-2-3 formula is

+E6/$E$15

Notice the [+] sign at the beginning to indicate a VALUE to 1-2-3. Remember that the cell addresses can be obtained by pointing, and that as they are obtained they can be made absolute if required by pressing function key [4/ABS] once. The Copy command is used to copy the formula down the column, and the column sum can then be calculated in F15.

To align the values correctly use the Range Format command, selecting either Fixed with 4 decimal places, or, as shown in table 5.8, Percent with 2 decimal places.

Table 5.8 Relative Frequencies, Cumulative Frequencies

	D	E	F	G	H
1					
2	------	------	------	------	------
3	UPPER CLASS		RELATIVE		CUM. REL.
4	LIMIT:	f	FREQ:	F	FREQ.
5					
6	5000	1	5.26%	1	5.26%
7	7500	9	47.37%	10	52.63%
8	10000	1	5.26%	11	57.89%
9	12500	3	15.79%	14	73.68%
10	15000	3	15.79%	17	89.47%
11	17500	1	5.26%	18	94.74%
12	20000	0	0.00%	18	94.74%
13	22500	1	5.26%	19	100.00%
14					
15		19	100.00%		

The **cumulative frequency**, F, of each class, is the sum of the frequencies of that class and of all previous classes.

That is,

$$F = \sum_{}^{\text{Current class}} f \qquad [5.2]$$

Each cumulative frequency, therefore, is the sum of the previous cumulative frequency and the frequency of the current class.

$$F = F_{previous} + f_{current} \qquad [5.3]$$

From a cumulative frequency distribution you can see directly how many values are not greater than the upper limit of each class. As with the frequency distributions from which they are constructed, this can be expressed either in absolute or in relative terms. We shall form both types of cumulative distribution from our two frequency distributors. Starting from our frequencies, **f**, in column E we shall form the cumulative frequencies, **F**, in column G. (Be careful to distinguish the column title, **F**, from the column in which the values are placed, which happens to be column G.) We shall also construct cumulative relative frequencies in column H.

Enter titles for columns G and H, then move the pointer to G6 where the first cumulative frequency is to be entered. This is the same as the frequency for the first class. Enter a formula to tell 1-2-3 that the value in G6 should be the same as in E6, that is, enter the formula

+E6

It is always better, where possible, to **use a formula** to obtain a new value **rather than a command** such as Copy. If values previously in the spreadsheet are changed, new values obtained from them by formulae will be recalculated automatically, while those entered by the Copy or Data commands will have to be updated by re-issuing these commands. The formula you have entered in G6 can be copied to H6, where it will appear as

+F6

You have copied a formula, not just a number, so this will bring across as the first cumulative relative frequency whatever the first relative frequency is.

Notice that it is possible to obtain the formula for the first cumulative relative frequency by copying because we are forming two columns of cumulative frequencies, each of which is in the same position (two columns to the right) with respect to its corresponding frequency column. In the same way, the formula we shall enter in G7 to calculate the cumulative frequency of the second class may be copied not just down that column, but also across to the corresponding cells of column H. Each cumulative frequency after the first is, according to formula 5.3, the sum of the previous cumulative frequency and the class frequency. We enter, therefore, in G7 the formula:

+G6+E7

Now use the Copy command as usual, but extend the range to Copy To so that it covers the cells for which values are required in both columns G and H. Your control panel will be as in table 5.9:

Table 5.9 Copying To a Rectangular Range

```
G7: +G6+E7                                                    POINT
Enter range copy FROM: G7..G7   Enter range copy TO: G7..H13
```

Pressing [RETURN] calculates and enters the figures. You can then use Range Format to display the cumulative relative frequencies as percentages and, after using Copy to extend the line drawn above the column headings, you should have the spreadsheet shown in table 5.8. Save a copy of the spreadsheet in this form, because we shall use it in the next chapter.

5.5 Graphs of Distributions

With the Graph command in 1-2-3 you can easily obtain pictures of your distributions. Frequency distributions can be displayed as histograms, called bar charts in 1-2-3, or alternatively, as frequency polygons using 1-2-3's XY graphs.

The half-arch shaped XY graph of a cumulative frequency distribution is called an ogive. If you print this using cumulative relative frequencies you can easily interpolate on it to estimate values for the median and quartiles.

If you would like to try producing these graphs, a few suggestions may be helpful in improving the picture you obtain. When drawing the cumulative frequency polygon it is natural to plot the cumulative frequencies on the Y axis (the 1-2-3 A range) against the upper class limits, which we have in column D, on the X axis. To ensure that the polygon touches the X axis you will need to insert an extra class at the bottom end of your distribution, which has a cumulative frequency of 0.

When plotting a frequency bar chart or polygon you should plot the class mid-points on the X axis. We shall construct a column containing these in chapter 6. To make your frequency polygon touch the X axis at both the lower and upper ends of the distribution you will have to insert additional classes at each end of the distribution, each with a frequency of 0.

5.6 Summary

We have now seen how to form frequency distributions and display them graphically. The larger your data set, the greater the benefit of 1-2-3's calculation and automatic recalculation facilities.

You should be gaining confidence in using formulae to perform calculations and in copying them as much as possible, sometimes to rectangular ranges. Remember to not just copy numbers in a spreadsheet but to use a formula to replicate them so that recalculation can take place.

We have learned some further commands for improving the appearance of our spreadsheets. We have learned to erase items in a specified range and to delete or insert entire columns or rows. Remember to save your spreadsheet before you erase or delete so that if you make a mistake you can retrieve the saved spreadsheet and try again.

6
Frequencies and Calculations

- **Printing spreadsheets**

6.1 Introduction

If you do not have the original data observations available but only a frequency distribution for the variable of interest, you may wish to **estimate** various **summary measures** for the data from the frequency distribution. We shall estimate the mean and standard deviation from the frequency distribution we constructed in chapter 5, so you should retrieve the spreadsheet shown in table 5.8.

In forming these estimates we shall use the frequencies as weights to be applied to the **X** values. The same computational method, then, could be used for calculating weighted means, such as index numbers.

You may wish to have a printed copy of all or part of your spreadsheet. This chapter details the use of the **Print** command which generates it.

6.2 Mean of Frequency Distribution

- **Worksheet Insert**

When individual values of a variable are not available you can estimate the mean, variance and standard deviation from a frequency distribution of the data. We shall calculate these estimates from the frequency distribution we constructed, beginning with that for the mean, which is defined as:

$$\text{Mean} = \sum fx / \sum f \qquad [6.1]$$

where **x** is the mid-point of each class of the distribution, **f** is the class frequency, and the summations are over all classes of the distribution.

To estimate the mean for our frequency distribution, then, we must begin by forming from our upper class limits a column of class mid-points which are the **x** values. The mid-point for each class lies halfway between the upper limits for the current and previous classes. If one of these limits is not defined you can give it a value following the pattern of spacing of the limits given.

You might like to insert your mid-points in the spreadsheet to the left of the upper class limits. To do this you first insert a blank column using the **Worksheet Insert** command. The column will be inserted immediately to the left of the current cell, so begin by positioning the pointer on column D. Obtain the menu and select, in succession, Worksheet, Insert and Column. The computer will respond, if D2 is your current cell, with

   Enter column insert range: D2..D2

Since we are inserting a column the row numbers in the range, although required, are irrelevant. If we wanted to insert more than one column we could stretch the range to the right, but for inserting just one column we can simply press [RETURN]. Entries which were previously in and to the right of column D are all moved one column to the right, and any cell references in formulae are adjusted accordingly. The spreadsheet will then contain a blank column.

Enter a title and the class midpoints using Data Fill. Since the class interval is 2500 and the upper limit of the first class is 5000, the first midpoint must be 3750. In your Fill command, therefore, you require a Start value of 3750 and a step of 2500. Notice that, although you may have defined the Fill range exactly, you cannot simply accept the offered Stop value of 20000 (if this is the last value you used), since this is smaller than the Stop value you require of 21250. When you have completed the command the class mid-points will be available. You can use them for graphing histograms, as suggested in the previous chapter. We shall now use them for estimating the mean and standard deviation of the distribution.

Our calculations will involve forming four new columns of values. For each we shall enter an appropriate formula at the top, then use the Copy command to obtain the remaining values in the column. To keep these columns close to the columns containing the mid points, **X**, and the frequencies, **f**, which are used in their calculation you may like to insert blank columns to use as working space. Alternatively, as we have done, you can save the spreadsheet in its current form and then delete the columns containing upper class limits, relative frequencies and cumulative frequencies, since these are not needed for the mean and standard deviation calculations.

Table 6.1 Mean and Standard Deviation Calculation

```
Soper & Lee                    Page 1                      23-Jan-87

            D       E       F         G         H             I
 1         ┌─────────────────────────────────────────────────────────
 2         │ ────────────────────────────────────────────────────────
 3         │ MID-
 4         │ POINT   f       fX      X-Mean   Sq(X-Mean)  f*Sq(X-Mean)
 5         │   X
 6         │  3750   1      3750    -5789.5   33518005.5   33518005.5
 7         │  6250   9     56250    -3289.5   10820637.1   97385734.1
 8         │  8750   1      8750     -789.5     623268.7     623268.7
 9         │ 11250   3     33750     1710.5    2925900.3    8777700.8
10         │ 13750   3     41250     4210.5   17728531.9   53185595.6
11         │ 16250   1     16250     6710.5   45031163.4   45031163.4
12         │ 18750   0         0     9210.5   84833795.0          0.0
13         │ 21250   1     21250    11710.5  137136426.6  137136426.6
14         │
15         │        19    181250                          375657894.7
16         │
17         │
18         │        Mean = Sum(fX)/Sum(f) =                   9539.47
19         │
20         │        Var = Sum(f*Sq(X-Mean))/Sum(f) =       19771468.14
21         │
22         │        Standard deviation = SQRT(Var) =           4446.51
```

As formula 6.1 shows, to estimate the mean we must multiply each **X** value by the corresponding frequency, **f**, giving an estimate of the total of the values in each class. These **fX** values are shown in column F of table 6.1. The formula entered in F6 is, therefore:

   +E6*D6

Relative addresses are appropriate here so that when the formula is copied down the column each **X** value is multiplied by its corresponding frequency.

The sum of these **fX** values, ΣfX, can be calculated in F15. It is an estimate of the overall total of the values in the distribution and so, as defined in formula 6.1, when we divide it by the number of observations, Σf, we obtain an estimate of the mean. This is calculated in cell I18 of table 6.1, using the formula

   +F15/E15

You need to enter an identifying label alongside I18, and you may like to give the name **Mean** to the cell itself, so that you can refer to it by name in formulae. You may find it interesting to compare your estimate of the mean with the median value we found earlier.

6.3 Standard Deviation Calculation

We shall now calculate the estimate for the **variance** of our data using the definitional formula:

   Variance = $\Sigma f (X - \text{Mean})^2 / \Sigma f$ [6.2]

and from this we shall obtain the standard deviation, since

   Standard deviation = $\sqrt{\text{Variance}}$ [6.3]

Table 6.1 shows the columns computed to calculate these measures. Column G contains the deviations of the **X** values from their mean, (**X**-**Mean**). Remember that the formula you enter at the top of this column should use a relative address for **X**, but an absolute one for the **Mean**. Column H contains the squares of the column G values, and column I holds the column H values each multiplied by the frequency, **f**, shown in column E. You will want to set a fixed format for these columns to align the decimal points, but some of the values computed are then too large to be displayed, so a row of asterisks is shown.

Use the Worksheet Column-width command and expand the columns with the [RIGHT] arrow key until all the values are displayed. Form the sum of the values in column I. Formula 6.2 shows that you obtain an estimate of the variance by dividing this sum by the sum of the column E values, Σf. Table 6.1 shows the variance calculated in cell I20 and its square root, the standard deviation,in cell I22. Remember to enter identifying labels. When you have completed your spreadsheet, save it for future use.

Notice that the inbuilt average and standard deviation functions which 1-2-3 offers, @AVG and @STD, cannot be used to estimate these values from a frequency distribution. But since, in this particular instance, you have the actual salary values available you may like to calculate from them the mean and standard deviation using the @ functions, and compare your results with the frequency distribution estimates.

6.4 Printing Spreadsheets

- **Print**

The **Print** command enables you to obtain a printed copy of your spreadsheet. If your computer has a printer attached, after calling up the menu you can select Print and then **Print Printer** to send your output directly to it. Print File is offered to you as an alternative. This is useful if you want to incorporate your spreadsheet in some other text, as we have done in this book. Most word processing packages can read a 1-2-3 print file, and you can then print your text and spreadsheet together. After you have chosen the destination of your output, the Print sub-menu will be displayed:

Table 6.2 The Print Sub-Menu

```
I23:                                                    POINT
|Range| Line Page Options Clear Align Go Quit
Specify a range to print
```

You must select **Range** and specify the area of the spreadsheet you wish to print: for the spreadsheet shown in table 6.1 this is D1 to I22. If the paper in your printer is not wide enough for the whole width of the range you specify to fit on it, 1-2-3 will split the spreadsheet for you, printing part of it below the rest. The appearance of your printout, however, may be improved if you select an appropriate section of the spreadsheet to print first, and follow this by printing another section.

After specifying a range you can select **Go** to start printing your spreadsheet.

If you then want to print another spreadsheet or spreadsheet section you may like to separate it from what you have just printed. You can select **Line** in the Print sub-menu to advance the printer one line, so doing this several times in succession will give you a gap of several lines. Alternatively you can select **Page** to advance the printer to the top of the next page. If you prefer to manually adjust the printer, leaving it set at the top of the next page, you must tell 1-2-3 that you have done this by selecting **Align**.

Selecting **Options** in the Print sub-menu enables you to specify in various ways how your spreadsheet is to be printed, and to add extra text to it. Table 6.3 shows the available commands.

Table 6.3 The Print Options Sub-Menu

```
I23                                                      POINT
Range Line Page |Options| Clear Align Go Quit
Header, Footer, Margins, Borders, Setup, Page-Length, Other
```

In printing table 6.1 the **Header** Option was used to print the top line of text. In response to the prompt: Enter Header Line: the words to be printed at the left, **Soper & Lee**, were typed, followed by the separator [!]. Preceding Header text by [!] causes it to be centred; a second [!] causes it to be right aligned. In this instance the first [!] was followed by **Page** and then the [#] sign, for which 1-2-3 substitutes the current page number. A second [!] sign was typed and then the [@] symbol. This symbol in a Header or Footer causes the current date, which you set as you turn on the computer, to be printed. **Footer** text can be set out in the same way as Header text. It is printed at the bottom of each page, but will only be printed on the last page when you select **Page**. **Margins** allows you to alter the Left, Right,Top and Bottom margins of your print page. The **Borders** Option can be used to print row and column headings. These must first be set up somewhere in the spreadsheet, and the ranges containing them specified as you use the option. The Borders command has been used to print the column letters and row numbers shown in table 6.1. **Setup** allows you to send control codes to your printer. You might use it, for example, to turn on compressed print so that you could set out more columns of your spreadsheet alongside one another. With the **Page-Length** option you can alter the default number of lines on a page. The **Other** option offers you the possibility of printing the cell formulae rather than the values they compute. This is useful in obtaining a record of how a spreadsheet has been constructed. Other also offers you the possibility of an "infinite" page length, so that page breaks, and also Headers and Footers, will never be printed. To leave the Print Options sub-menu you must select **Quit**, which returns you to the Print menu.

The Print **Clear** command cancels some or all of the printing instructions you have set. When you have finished printing, or if you want to make an alteration to your spreadsheet, select **Quit** to return to READY mode.

6.5 Summary

In this chapter we have calculated summary measures for a frequency distribution using definitional formulae.

We have learned the 1-2-3 commands for inserting entire columns or rows and for printing spreadsheets. Try out these commands. Alter the layout of your worksheet as you wish, then Print it in suitable sections, each with identifying Borders. Use the Header and Footer facilities to title, date and page number your work.

7
Probability

- **Database facilities**
- **Changing a formula to a number**

7.1 Measuring Probability

We measure the **probability** of an **event** as a number between 0 and 1. An impossible event has probability 0. If an event is certain to occur, perhaps because of the way it is defined, then it has probability 1. Most events are neither impossible nor certain, but somewhere between the two. Their probability is a fraction between 0 and 1, which may be displayed in a spreadsheet either as a decimal or as a percentage.

When we select one person at random from a list, the probability that that person has a particular attribute, A, is the **relative frequency** with which that attribute occurs amongst people on the list. We have

$$P(\mathbf{A}) = \frac{\text{number of persons with attribute } \mathbf{A}}{\text{total number of persons}} = \frac{f}{N} \qquad [7.1]$$

where **f** is the frequency with which attribute **A** occurs and **N** is the total number of people on our list.

The employee personal data spreadsheet shown in table 7.1 is an extended version of the one we used in chapter 5. More employees have been added, and we now have listed for each employee the person's sex, whether full or part time hours are worked, and an identification number. You should set up this new spreadsheet, by retrieving and modifying your earlier one.

Table 7.1 Further Employee Personal Data

	A	B	C	D	E	F
1	EMPLOYEE PERSONAL DATA					
2	--------	---------	-------	----	-----	----
3	SURNAME:	FORENAME:	SALARY:	SEX:	TIME:	ID:
4	Smith	Angela	4800	F	PART	1
5	Jones	Roberta	5110	F	PART	2
6	Reed	Jane	5520	F	PART	3
7	Mackie	Charles	5570	M	PART	4
8	Soper	Jean	6325	F	PART	5
9	Archer	Sean	6750	M	PART	6
10	Phillips	John	6785	M	PART	7
11	Evans	Paula	7320	F	PART	8
12	Lawson	Hugh	7320	M	PART	9
13	Smith	Julia	7320	F	PART	10
14	Thatcher	Edward	8894	M	PART	11
15	Lee	Martin	10655	M	FULL	12
16	Young	Peter	10760	M	FULL	13
17	White	Lee	12480	M	FULL	14
18	Cox	Henry	13560	M	FULL	15
19	Baker	Rachel	14380	F	FULL	16
20	Bingham	John	14380	M	FULL	17
21	Brown	John	15460	M	FULL	18
22	McKay	Iain	20860	M	FULL	19
23	Jeffreys	Sue	9500	F	PART	20
24	Quarmby	Rhoda	10500	F	FULL	21
25	Macklin	Jackie	11500	F	FULL	22
26	Ralhan	Veena	12500	F	FULL	23
27	Palin	Petrina	13500	F	FULL	24
28	Andrews	Derek	14500	M	FULL	25
29	Peacock	Derek	15500	M	FULL	26

In this chapter we shall calculate the probability that a person selected at random from this list, or from a particular part of it, has a certain attribute or attributes. The spreadsheet approach shows us exactly how each probability is formed, because in each case we can generate a separate list of the persons satisfying the criterion and then count them. We shall calculate, for example, the probability that a person is male or works full time, and we shall avoid any possibility of double counting persons who are both. In this way we shall illustrate the various rules of probability.

The Data Query command in 1-2-3 allows us to define a particular set of data as constituting a **database**. The command can then further be used to extract the records of people with particular attributes from that database. These can then be counted to find f. The @COUNT function can be used to do this, and also to count the total number, **N** on our list so that the probability defined in 7.1. can be calculated. Later we shall see how to use the special @DCOUNT function for counting records which satisfy a particular criterion directly from the database.

We shall begin by looking at some of the database facilities of 1-2-3 and finding out about the Data Query command. We shall then go on to utilise it in the calculation of probabilities.

7.2 Database Facilities

- **Data Query**
- **Input, Criterion and Output ranges**
- **Finding and Extracting**

The **Data Query** command in 1-2-3 is used in conjunction with a set of **records**, such as our employee personal data, which have been entered in the spreadsheet in the usual way and are then defined to constitute a 1-2-3 **database**. Each row of data in the database consitutes a record, in our case the information about a particular employee. The database must also contain, **immediately above** the first data row, a row of titles, or **field names** for each column. In table 5.4, to improve readability, we had a row containing only the units of measurement, (ú), between the column titles and the first data record. Before defining a database this row must be removed, as shown in table 7.1.

It is important **not** to use **spaces** at the beginning or end either of entries in the database, or of field names. They would be invisible to you and would cause problems when using data commands which work by exactly matching specified items with ones in the database.

It is convenient to use the Range Name Create command to give the name **DB** to the column titles and associated entries which comprise the database. You must then define **DB** to be the database by specifying it as the **Input** range. To do this, select Data Query Input from successive menus and give **DB** as the requested Input range.

To be able to select people with certain attributes from the database we need a **Criterion** range which specifies the attribute or attributes required. To set this up, Quit from the Data Query sub-menu. You can then enter information to define your Criterion range in an empty section of your spreadsheet. The first row of your Criterion range must contain, in identical form, at least some of the field names which are at the top of your Input range. The easiest way of ensuring they are identical is to use the Copy command to replicate them.

Then below the appropriate field name(s) you enter, in **exactly** the same form as in the database, the attribute(s) or value(s) which you wish to match. Begin by entering **Jones** immediately below the field name SURNAME:. Notice that in the Data Query command 1-2-3 distinguishes capital letters from lower case ones. It also checks whether spaces and punctuation marks match. The Criterion range is shown in table 7.2.

Table 7.2 Criterion Range

	A	B	C	D	E	F
32	SURNAME:	FORENAME:	SALARY:	SEX:	TIME:	ID:
33	Jones					

Now you have set up a Criterion range you must define it as such. Again choose Data from the menu, followed by Query and now Criterion. You will be asked to

    Enter Criterion range:

and should specify the range shown in table 7.2 which is two rows deep by six columns wide.

You may now select **Find** from the Data Query sub-menu. It will highlight in the database the first record which matches the criterion set, that is, the first person whose surname is Jones. She is Roberta Jones, the second person listed. Press [DOWN] to move to the next person whose surname is Jones. But the computer will "beep" and the screen will not change, since there is no-one else with that surname. Press [RETURN] to return to the Data Query sub- menu. Quit from it and try entering a different surname, say **Smith**, in place of **Jones** in the Criterion range. Try using Data Query Find again. You should find two persons who satisfy this criterion, Angela Smith and Julia Smith who are 1st and 10th respectively in the database list.

To form a list of the people who satisfy a certain criterion we must specify an **Output** range into which their records can be placed. To set up field names for an Output range, [RETURN] from using Find, Quit from Data Query and again use the Copy command to replicate the field names in a section of the spreadsheet which has empty space beneath it. Now return to the Data Query command and this time select Output. You should define the Output range to be simply the row of field names you have just entered. 1-2-3 will then use as many rows as required beneath these field names to hold copies of the records you select.

Your Input and Criterion ranges remain as you previously defined them. You may now choose Extract from the sub-menu and, if you have entered **Smith** as your criterion, the records of the two Smiths will appear in your Output range. Note that to view them you may have to Quit Data Query so that you can move the pointer to the appropriate part of the screen. Table

7.3 shows the Output range you should have if you copied all the field names to it. It is possible to use just some of the field names in the Output range. Only the parts of the records corresponding to these fields will then be copied.

Table 7.3 Data Extracted into Output Range

	A	B	C	D	E	F
39	SURNAME:	FORENAME:	SALARY:	SEX:	TIME:	ID:
40	Smith	Angela	4800	F	PART	1
41	Smith	Julia	7320	F	PART	10

You might like to try removing some of your Output field names using Range Erase, then using Data Query Extract again. You will find that only selected parts of your records have now been copied.

Now that we know how to define the various Data Query ranges we shall see how to count data records and will then proceed with calculating probabilities. In the process we shall find ourselves using the various Data Query ranges in different ways. Since we shall begin by altering the spreadsheet entries, Quit now from Data Query.

7.3 Counting Data Records

- **Output range defined as max. size required**
- **Criterion range containing blank row**

We can count the number of records in our database just as we usually count the number of items in a column, that is, by using @COUNT. If you Range Name the column of the database which contains the surname entries **PERSONS IN DB** you can count those persons by entering the formula @COUNT(**PERSONS IN DB**) as shown in table 7.4. The value thus computed is named **N**.

The Criterion range in table 7.4 comprises the field names SURNAME: SEX: and TIME:, since these are the criteria we are concerned with, together with the row of cells below, which contains the entry **Smith**.

The field names for the Output range in table 7.4 are SURNAME: FORENAME: SEX: TIME: and ID: Previously we simply defined such labels as the Output range. So that we can make further use of the entries in the Output range, however, it is now advantageous to define the Output range to contain **all the rows that might be required** for copying into from the input range.

One way of seeing how many rows are needed is to use the Data Query Extract command with a **blank row in the Criterion range**. This extracts all records in the Input range. You can include a blank row in the Criterion range either by Range Erasing existing entries, or by redefining the Criterion range by extending it downwards to include a blank row. The Output range is defined for the moment as just the appropriate field names. The Extract command will then extract information from all your database records, placing it immediately below the Output field names. You can then redefine the Output range to include both the field names and all the extracted data, and you can Range Name it **OUT**.

Let us set up an @COUNT function to count all the records in this Output range. When we first do this, say by range naming the column containing the surnames **PERSONS IN OUT** and entering @COUNT(**PERSONS IN OUT**), we will of course obtain the same value as in our earlier count of the database. @COUNT, however, counts only non-blank cells. Alter your Criterion range so it does not include blank rows and **Smith** is entered under SURNAME:

Use Data Query Extract again. The Output area of your spreadsheet will alter to show just two entries. Notice that @COUNT does not immediately recalculate, but it will do so as you Quit from Data Query and will then show a count of 2 as in table 7.4, where it is named **f**.

Table 7.4 Persons Counted to find Probability

```
        A          B           C          D          E
32   SURNAME:   SEX:        TIME:              @COUNT(PERSONS IN DB)
33   Smith                                            26
34                                                     N
35              PROBABILITY                    @COUNT(PERSONS IN OUT)
36                 f/N                                 2
37                0.077                                f
38
39   SURNAME:   FORENAME:SEX:           TIME:       ID:
40   Smith      Angela     F            PART         1
41   Smith      Julia      F            PART        10
```

We have defined the probability of selecting at random a person with a particular attribute, **A**, to be:

$$P(\mathbf{A}) = \frac{\text{number of persons with attribute } \mathbf{A}}{\text{total number of persons}} = \frac{f}{N} \qquad [7.1]$$

We have now set up counts to find the values of **f** and **N** so that we can find P(**A**) by calculating their quotient. In our spreadsheet the attribute **A** is what is entered in the Criterion range, which in table 7.4 is **Smith**. From the cell labelled **PROBABILITY**, then, we can see that the probability of a person who is selected at random from the list having the surname Smith P(**Smith**) is 0.077. Save your completed spreadsheet.

You may then like to try finding the probability of selecting a person with a different attribute, say one who works full time. Alter your Criterion range appropriately, by Range Erasing the unwanted entry, **Smith**, and by entering **FULL** under the heading TIME: Again use the Data Query Extract commands, then quit the menu so that recalculations can take place. You should find that **f** is now 14, which is the number of persons who work full time, and **f/N**, the probability of choosing such a person, P(**FULL**), is 0.538. By repeating this procedure you can find other probabilities. For example, the probability of randomly selecting a person whose surname is Jones, P(**Jones**), is 0.038.

7.4 Probability Rules

- **OR Criteria**
- **AND Criteria**
- **@DCOUNT**
- **[2/EDIT] [9/CALC]**

The addition rule for mutually exclusive events is taken as one of the axioms, or self-evident truths, of probability. It allows us to add the separate probabilities of events which cannot simultaneously occur to find the probability that one of them occurs. We have:

$$P(\mathbf{A} \text{ or } \mathbf{B} \text{ or } ...) = P(\mathbf{A}) + P(\mathbf{B}) + ... \qquad [7.2]$$

No person can have more than one surname so we can use this rule to find, in our database, the probability that a person is called Smith or Jones. We have:

P(**Smith** or **Jones**) = P(**Smith**) + P(**Jones**)

Substituting our earlier values for the right hand side probabilities gives

P(**Smith** or **Jones**) = 0.077 + 0.038 = 0.115

With 1-2-3, however, we can also calculate the left hand side probability directly because we can set up as our criteria one attribute (**Smith**) or another (**Jones**) by entering them on different lines in the criterion range. Table 7.5 shows a Criterion range containing the entries **Jones** and **Smith**, an Output range with the appropriate records extracted and P(**Jones** or **Smith**) calculated as 0.115.

Table 7.5 "OR" Criterion, Mutually Exclusive Events

	A	B	C	D	E
32	SURNAME:	SEX:	TIME:		@COUNT(PERSONS IN DB)
33	Jones				26
34	Smith				N
35		PROBABILITY			@COUNT(PERSONS IN OUT)
36		f/N			3
37		0.115			f
38					
39	SURNAME:	FORENAME:	SEX:	TIME:	ID:
40	Smith	Angela	F	PART	1
41	Jones	Roberta	F	PART	2
42	Smith	Julia	F	PART	10

This spreadsheet is obtained by following our earlier procedure for finding a new probability, but notice that once the entries in the Criterion range are correct this range must be redefined to include both entries in it before the Extract command is used.

Table 7.6 "OR" Criterion, Non-Mutually Exclusive Events

	A	B	C	D	E
32	SURNAME:	SEX:	TIME:		@COUNT(PERSONS IN DB)
33			FULL		26
34		M			N
35		PROBABILITY			@COUNT(PERSONS IN OUT)
36		f/N			19
37		0.731			f
38					
39	SURNAME:	FORENAME:	SEX:	TIME:	ID:
40	Mackie	Charles	M	PART	4
41	Archer	Sean	M	PART	6
42	Phillips	John	M	PART	7
43	Lawson	Hugh	M	PART	9
44	Thatcher	Edward	M	PART	11
45	Lee	Martin	M	FULL	12
46	Young	Peter	M	FULL	13
47	White	Lee	M	FULL	14
48	Cox	Henry	M	FULL	15
49	Baker	Rachel	F	FULL	16
50	Bingham	John	M	FULL	17
51	Brown	John	M	FULL	18
52	McKay	Iain	M	FULL	19
53	Quarmby	Rhoda	F	FULL	21
54	Macklin	Jackie	F	FULL	22
55	Ralhan	Veena	F	FULL	23
56	Palin	Petrina	F	FULL	24
57	Andrews	Derek	M	FULL	25
58	Peacock	Derek	M	FULL	26

Certain events are not mutually exclusive. They may both (or, more generally, all) occur together. The **Addition Rule** for **two non-mutually exclusive events** is:

$$P(\mathbf{A} \text{ or } \mathbf{B}) = P(\mathbf{A}) + P(\mathbf{B}) - P(\mathbf{A} \text{ and } \mathbf{B}) \qquad [7.3]$$

For the right hand side of this expression we require P(**A** and **B**) which is the probability that both events **A** and **B** occur, or the **joint probability** of these events. The left hand side of 7.3, however, can be directly calculated with 1-2-3, in the same way as for mutually exclusive events.

Let us find the probability of randomly selecting a person who works full time or who is male, P(**FULL** or **M**). Certain persons on our list satisfy both criteria, but if **FULL** and **M** are placed in different rows of the Criterion range, as shown in table 7.6, the Data Query Extract command will merely extract the records of all persons who satisfy at least one of the criteria. Thus there is no problem of double counting, and as we Quit the Data Query menu the required probability, P(**FULL** or **M**) is calculated and displayed as 0.731.

Let us now find the joint probability that a person chosen from our database works full time **and** is male. We require P(**FULL** and **M**) which we can find directly in 1-2-3 by entering both criteria in the same line of our criterion range and issuing appropriate commands. Remember to redefine your Criterion range so that it does not include a blank row. The Output you should Extract is shown in table 7.7. Notice that it consists of the records of those people who satisfy both the criteria set. The recalculated probability value of 0.346 is the probability that a randomly chosen person is a male full time worker, denoted P(**FULL** and **M**).

The probability that a person works full time may depend on that person's sex. Thus we may define the conditional probability P(**FULL/M**) to be the probability that a man works full time. We can find this probability by considering only the men in our database, that is, those who fulfil the condition set, and counting the number of these who work full time. In more general notation we require P(**A/B**) where **A** is the probability that a person works full time, and **B** is that the person is male. We shall demonstrate the result that

$$P(\mathbf{A/B}) = P(\mathbf{A} \text{ and } \mathbf{B}) / P(\mathbf{B}) \qquad [7.4]$$

by finding both sides of the expression

$$P(\mathbf{FULL/M}) = P(\mathbf{FULL} \text{ and } \mathbf{M}) / P(\mathbf{M}) \qquad [7.5]$$

and showing that they are equivalent.

Table 7.7 "AND" Criterion

	A	B	C	D	E
39	SURNAME:	SEX:	TIME:		@COUNT(PERSONS IN DB)
40		M		FULL		26
41					N
42		PROBABILITY			@COUNT(PERSONS IN OUT)
43		f/N			9
44		0.346			f
45					
46	SURNAME:	FORENAME:	SEX:	TIME:	ID:
47	Lee	Martin	M	FULL	12
48	Young	Peter	M	FULL	13
49	White	Lee	M	FULL	14
50	Cox	Henry	M	FULL	15
51	Bingham	John	M	FULL	17
52	Brown	John	M	FULL	18
53	McKay	Iain	M	FULL	19
54	Andrews	Derek	M	FULL	25
55	Peacock	Derek	M	FULL	26

The calculations we shall make are shown in table 7.8. We have found P(**FULL** and **M**) and would like to retain it while we calculate our other probabilities. Use the Copy command to place a copy of this probability in some convenient cell. By doing this you actually copy the formula used to calculate the probability. You must then **convert** the copied **formula into a numerical value**, so that it will not recalculate as we compute other probabilities. To do this with 1-2-3 release 1, position the pointer on the cell to which you copied the probability and press [2/EDIT] followed by [9/CALC]. This evaluates the formula and replaces it by its numerical result. Look at the cell contents displayed in the control panel to check that this is so. Remember to label your copied value. An alternative method, using a 1-2-3 release 2 command, is given in Chapter 18.

Now set the Criterion range to be just the two cells containing the field name **SEX:** and the entry **M** and use the Extract command. You will extract all the men in the database, and on Quitting Data Query the probability P(**M**) will be calculated as 0.538. You can then evaluate the right hand side of equation 7.5, P(**FULL** and **M**) / P(**M**), which is 0.643.

To find the conditional probability P(**FULL/M**), since we know the person we are concerned with is a male, we need consider only the men in our database, or the persons who are listed in the Output range of table 7.8. The proportion of these who work full time is our required probability:

$$P(\textbf{FULL/M}) = \frac{\text{no. of full time men}}{\text{total number of men}} = \frac{\textbf{fFULL}}{f} \qquad [7.6]$$

We need to count how many people in the Output range satisfy the Criterion **FULL**. Now that we are familiar with Input and Criterion ranges we shall use the @DCOUNT function to do this.

Table 7.8 Calculation of Conditional Probability

```
       B           C           D          E           F
32  SEX:        TIME:                  @COUNT(PERSONS IN DB)
33  M           FULL                       26
34                                         N
35  PROBABILITY                        @COUNT(PERSONS IN OUT)
36      f/N                                14
37      0.538                              f  P(FULL and M)
38                                            0.346
39  FORENAME:SEX:           TIME:      ID:
40  Charles     M           PART        4  P(FULL and M)/P(M)
41  Sean        M           PART        6     0.643
42  John        M           PART        7
43  Hugh        M           PART        9
44  Edward      M           PART       11 @DCOUNT(OUT,0,CRITF)
45  Martin      M           FULL       12        9
46  Peter       M           FULL       13     fFULL
47  Lee         M           FULL       14
48  Henry       M           FULL       15   P(FULL/M)
49  John        M           FULL       17     fFULL/f
50  John        M           FULL       18     0.643
51  Iain        M           FULL       19
52  Derek       M           FULL       25
53  Derek       M           FULL       26
```

The **@DCOUNT** function is one of a number of statistical functions designed for database use and all beginning with @D. Each of these functions takes the form

>   @function name(Input range,Offset,Criterion range)

The input and criterion ranges are specified as those we have become accustomed to for the Query command. We now want to use our previous Output range, which you should have named **OUT**, as the @DCOUNT Input range. The Criterion range for our function is the two cells named **CRITF** which contain the field name **TIME:** and the entry **FULL**. The offset tells 1-2-3 which column of the Input range it should use in its counting. The left hand column is offset 0, the next to the right is offset 1, and so on. Since there are no partly filled rows in our new input range it does not matter which column of items we count, so we will use offset 0 and count the surnames.

Choose a vacant space in your spreadsheet and enter the function

>   @DCOUNT(**OUT**,0,**CRITF**)

together with an appropriate label. Range Name the value **fFULL**. The total number of men is already calculated at 14 and named f, so you can evaluate the conditional probability of equation 7.6, P(**FULL/M**), by forming the quotient **fFULL/f**, which has the value 0.643. Notice that this is the same as the value we calculated earlier for the right hand side of 7.5.

Comparing the conditional probability P(**FULL/M**) with the probability P(**FULL**) which we found in section 7.3 to be 0.538, we notice that they are not equal. This is because in the data we are dealing with, different proportions of men and of women work full time. We say that a person's sex and his or her work pattern are **not independent**. If events **A** and **B** are independent,

$$P(\mathbf{A}) = P(\mathbf{A/B}) \qquad [7.7]$$

Rewriting our expression for conditional probability, 7.4, gives the **Multiplication Rule**

$$P(\mathbf{A} \text{ and } \mathbf{B}) = P(\mathbf{B}).P(\mathbf{A/B}) \qquad [7.8]$$

and we may alternatively write

$$P(\mathbf{A} \text{ and } \mathbf{B}) = P(\mathbf{A}).P(\mathbf{B/A}) \qquad [7.9]$$

Equating the right hand sides of 7.8 and 7.9 and re-arranging gives Bayes Theorem:

$$P(\mathbf{B/A}) = \frac{P(\mathbf{B}).P(\mathbf{A/B})}{P(\mathbf{A})} \qquad [7.10]$$

In our context this defines what is often called the **posterior probability** that a person chosen at random is a man, given that we know it is someone who works full time. We can write

$$P(\mathbf{M/FULL}) = \frac{P(\mathbf{M}).P(\mathbf{FULL/M})}{P(\mathbf{FULL})} \qquad [7.11]$$

We have already calculated the various probabilities on the right hand side of expression 7.11. You might like to calculate the left hand side also.

Sometimes we do not have information to work out the denominator of 7.10, P(**A**), directly, but we can evaluate it by considering the various mutually exclusive events in conjunction with which event **A** occurs. In the case we have been considering, a full time worker must be either male or female.

So we have:

P(**FULL**) = P(**FULL** and **M**) + P(**FULL** and **F**)

You might like to verify this expression.

Using the more general notation, if event **A** occurs either in conjunction with event **B** or with its complement, denoted **B'** we would write

P(**A**) = P(**A** and **B**) + P(**A** and **B'**)

Applying the Multiplication Rule of equation 7.8 to each term in the right hand side gives

P(**A**) = P(**B**).P(**A/B**) + P(**B'**).P(**A/B'**)

which allows us to rewrite equation 7.10 as

$$P(\mathbf{B/A}) = \frac{P(\mathbf{B}).P(\mathbf{A/B})}{P(\mathbf{B}).P(\mathbf{A/B})+P(\mathbf{B'}).P(\mathbf{A/B'})} \qquad [7.12]$$

This form of expression for a posterior probability is used in decision analysis.

7.5 Summary

In this chapter we have used the database facilities of 1-2-3. We have learned how to define the Input, Criterion and Output Data ranges. We have used Find and Extract from the Data Query command, and have counted database records using the @DCOUNT function.

Using a spreadsheet in this way has enabled us to see exactly what we are counting when we evaluate the probabilities of various events.

8
Probability Distributions

- @VLOOKUP
- Graph Options Scale
- @PI, @EXP

8.1 Introduction

Probability distributions are the theoretical counterpart to frequency distributions. If the variable, **X**, which we are measuring can take on only certain specific values it is a **discrete** variable, and its probability distribution gives the probability of each of the **X** values occurring. The **binomial** distribution is one such distribution, and we shall calculate probabilities for it from a general formula.

Their computation will involve us in creating a table of values and looking up the ones which we require using @VLOOKUP. We shall graph our distributions and find that 1-2-3's automatic recalculation facility makes it easy for us to study the effects of changes.

When **X** can take any value in a certain range it is a **continuous** variable, and its probability distribution is described by what is called a **probability density function**. We can evaluate the numerical probability of **X**, say, being greater than a certain value, using the mathematical technique of integration. This measures the area under the probability density function above the **X** value specified, and this area represents the desired probability. With a computer it is necessary to use an approximation to do this. A method of calculating such a probability is given in section 4 of this chapter for the **normal, t, Chi² **and **F** distributions, which are commonly used in statistical inference.

Prior to that, in section 3, we shall graph the density function for one of these distributions, the normal, and we shall see how the graph alters if we change the specification of the function by altering its mean and standard deviation. To display such changes effectively we shall have to use the manual option for setting graph scales.

8.2 Binomial Distribution

- @VLOOKUP

When we are dealing with a series of events, each of which falls into one of two categories, the binomial distribution may apply. For example, suppose a random sample of items are classified as either good or defective. If the sample size is small relative to that of the population, the probabilities of the various possible numbers of defectives which may occur can be calculated by the binomial formula.

Let **n** be the sample size, and **p** the proportion of defectives in the population. Then the number of defectives, **X**, in the sample must be between 0 and **n**, and for each possible value of **X** we can calculate its probability by the formula

$$P(X) = {}^nC_X \, p^X \, (1-p)^{n-X} \qquad [8.1]$$

where ${}^nC_X$ is the number of combinations of **X** items from **n**, defined as

$$ {}^nC_X = n!/[X!(n-X)!] \qquad [8.2]$$

and where **X!** (read as **X** factorial) is defined as

$$X! = X \, (X-1) \, (X-2) \, \ldots \ldots \, 2 \, . \, 1 \qquad [8.3]$$

For example, if a random sample of 7 cups are chosen from a crate in which 20% are defective, what is the probability that in the sample 3 cups are defective? We have n = 7, p = 0.2 and X = 3. Substituting these values into the binomial formula, 8.1, gives

$$P(3) = {}^7C_3 \, 0.2^3 \, (1 - 0.2)^{7-3}$$

$$= 7!/(3! \, 4!) \, 0.2^3 \, 0.8^4 = 0.115 \qquad [8.4]$$

Table 8.1 Calculation of Factorials

	A	B
1	Calculations for Binomial Probabilities	
2		
3	Value	Factorial
4	0	1
5	1	1
6	2	2
7	3	6
8	4	24
9	5	120
10	6	720
11	7	5040
12	8	40320
13	9	362880
14	10	3628800
15	11	39916800
16	12	4.8E+08
17	13	6.2E+09
18	14	8.7E+10
19	15	1.3E+12
20	16	2.1E+13
21	17	3.6E+14
22	18	6.4E+15
23	19	1.2E+17
24	20	2.4E+18

Let us set up a spreadsheet to estimate the probability of each of the possible values of **X** when **n** = 7 and **p** = 0.2. The formula shown in 8.1 involves the factorials specified in 8.2. From the definition of X! given in 8.3 we can see that **X**! = **X** (**X** - 1)!. Because of this it is easiest to calculate factorials for a sequence of numbers, set them in a table, and then select those which are needed for a particular probability calculation. The largest factorial we require is n!, so to allow for a sample size of up to 20 we set up the values and their corresponding factorials shown in table 8.1. Use Data Fill to set up the column of Values, then calculate the corresponding factorials in the column to the right. 0! is 1 by definition, so enter 1 at the top of the Factorial column. Below that enter a formula to multiply the value to the left by the previous factorial, so that for **X** = 1 you form **X***(**X**-1)!. A numerical value of 1 is obtained. Copying the formula down the column evelutes all the required factorials. Ensure you have titled your columns, and Range Name **FACT** the values in the two columns comprising the table of factorials.

To evaluate ^nC_X we enter **n** and, below it in the same column, all the possible values of **X**, together with appropriate labelling as in table 8:2. If you Range Name **n** the cell in which you have entered that value, you can obtain the **X** values using Data Fill and entering **n** as the Stop value. Next form the column headed **n-X**. Remember to use an absolute address for **n** in your formula so that it can be copied.

Table 8.2 Calculation of the Binomial Coefficients

	C	D	E	F	G
3			@VLOOKUP(n,$FACT,1)		
4		n	n!		
5		7	5040		
6					
7	X	n-X	X!	(n-X)!	nCX
8	0	7	1	5040	1
9	1	6	1	720	7
10	2	5	2	120	21
11	3	4	6	24	35
12	4	3	24	6	35
13	5	2	120	2	21
14	6	1	720	1	7
15	7	0	5040	1	1

Binomial Distribution
n = 7, p = 0.2

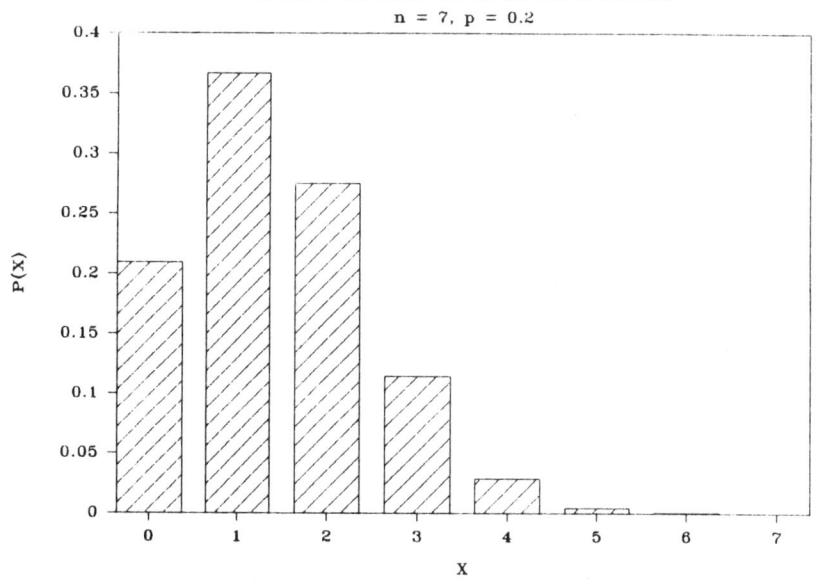

Fig. 8.1 Binomial Distribution, n=7, p=0.2.

We shall now find the factorials corresponding to **n** and to the various values listed for **X** and for **n-X**. We do this by using the @VLOOKUP function to look up their values in the table we named **FACT**. To find **n!** we enter, two columns to the right of n, the formula:

@VLOOKUP(**n**,**$FACT**,1)   [8.5]

The second argument in the parentheses defines the look up range to be the table we have Range Named **FACT**. Preceding the name with a [$] sign makes the range address absolute, which is necessary if the formula is to be copied. @VLOOKUP looks up the first value in the parentheses, **n** in this case, in the left hand column of the look up range, that is, in the column titled **Value**. The values in this column must be in ascending size order. The function then moves right the number of columns specified by the third argument, 1 in this case, and the value it locates there is displayed as the value of the function. So for **n** = 7 we look up the value 5040, which is 7! . The factorials looked up in table 8.1 are all displayed two columns to the right of the corresponding value of **n, X** or **n-X**. Because of this, formula 8.5 which has been entered for **n** can be copied to the appropriate rectangular range where it will look up the other factorials. Notice that the formula uses a relative address for the value whose factorial is being looked up, and an absolute address for the look up range.

Binomial Distribution
n = 7, p = 0.5

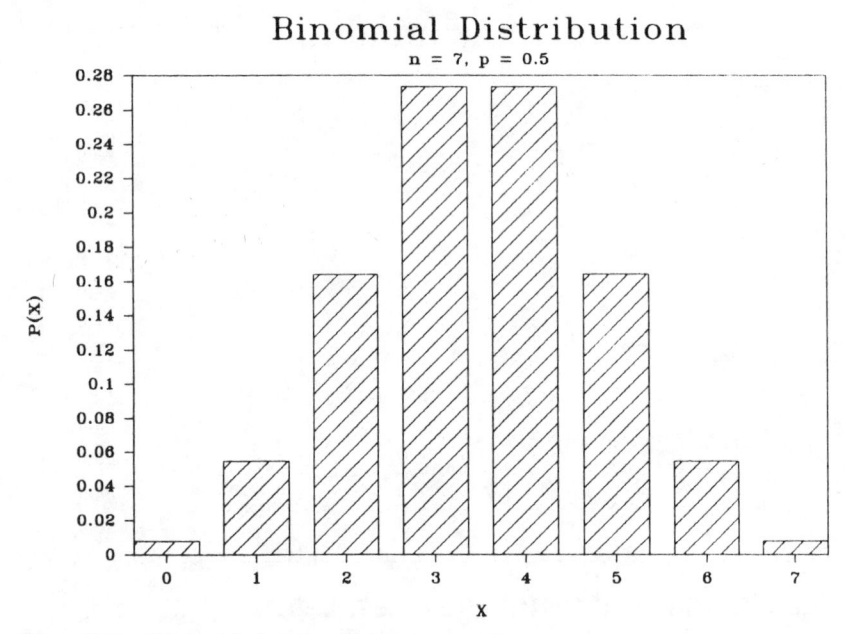

Fig. 8.2 Binomial Distribution, n=7, p=0.5.

The values of ^nC_X can now be found. They are shown in the column titled **nCX**, and are obtained by dividing **n!**, which needs an absolute address, by **X!** and by **(n-X)!**. Relative addresses for the latter two factorials will enable the formula to be copied down the column.

Formula 8.1 shows that the probabilities, P(**X**), which we require are calculated by multiplying the **binomial coefficients**, ^nC_X, which we have just found, by **p^X*(1-p)^(n-X)**. Enter and Range Name a value for **p**. In our example it is 0.2. Use this to calculate a column of values of **p^X**, using cell addresses for the **X** values as before. Compute 1-**p** followed by a column of values of (1-**p**)^(**n-X**). The products of the values in the three columns headed **nCX**, **p^X** and **(1-p)^(n-X)** in table 8.3, then, give the values of P(**X**), which are also shown.

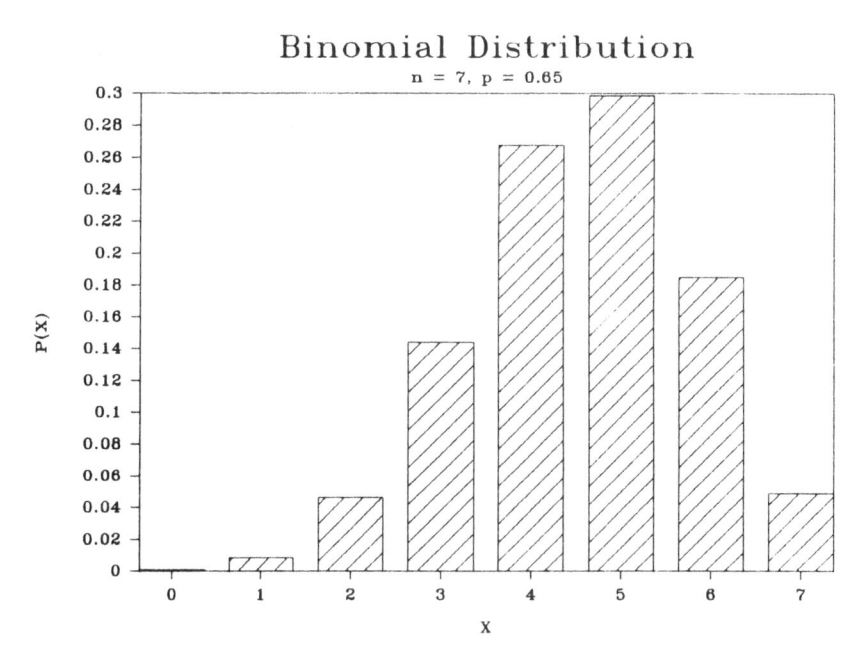

Fig. 8.3 Binomial Distribution, n=7, p=0.65.

We have found the values of the probability distribution for the binomial variable **X** when **n** = 7 and **p** = 0.2. Check that when **X** = 3 the probability is the same as that calculated in equation 8.4. Check also that the probabilities sum to 1. The distribution can be displayed as a Bar graph as shown in fig 8.1.

Table 8.3 Binomial Probability of X Successes in n Trials

	G	H	I	J
4			p	1-p
5			0.2	0.8
6				
7	nCX	p^X	(1-p)^(n-X)	P(X)
8	1	1	0.2097152	0.209715
9	7	0.2	0.262144	0.367001
10	21	0.04	0.32768	0.275251
11	35	0.008	0.4096	0.114688
12	35	0.0016	0.512	0.028672
13	21	0.00032	0.64	0.004300
14	7	0.000064	0.8	0.000358
15	1	0.000012	1	0.000012
16				
17				1.000000

Binomial Distribution
n = 20, p = 0.2

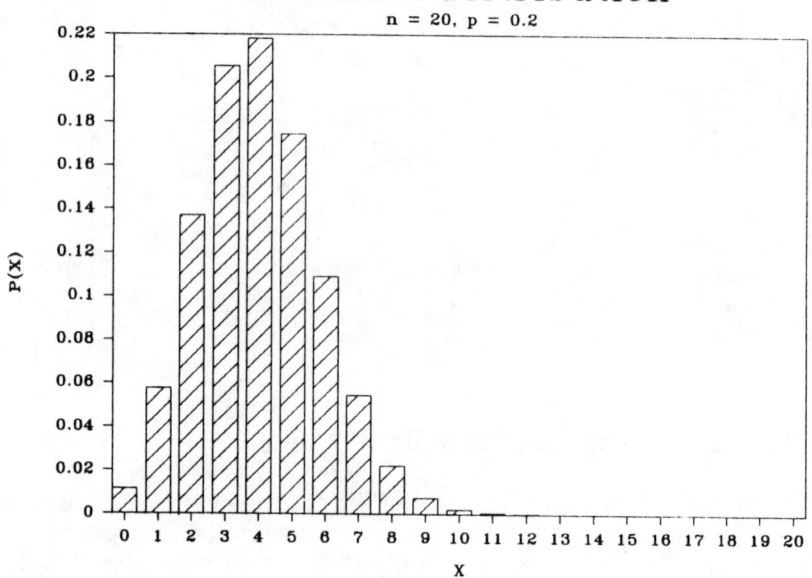

Fig. 8.4 Binomial Distribution, n=20, p=0.2.

You can obtain probabilities for other values of **n** and **p** simply by re-entering appropriate values and extending the columns by copying if a larger **n** is used. Try changing **p** to 0.5 and then to 0.65. Watch the shape of the graph change, as shown in figs 8.2 and 8.3.

For **n** = 20 and **p** = 0.2 you should obtain the bar chart shown in fig 8.4. Notice that even with **p** not equal to 0.5 the distribution becomes more symmetric as **n** increases. It can be proved theoretically to approach the normal distribution, which we shall graph in the next section.

8.3 The Normal Distribution

- @PI
- @EXP

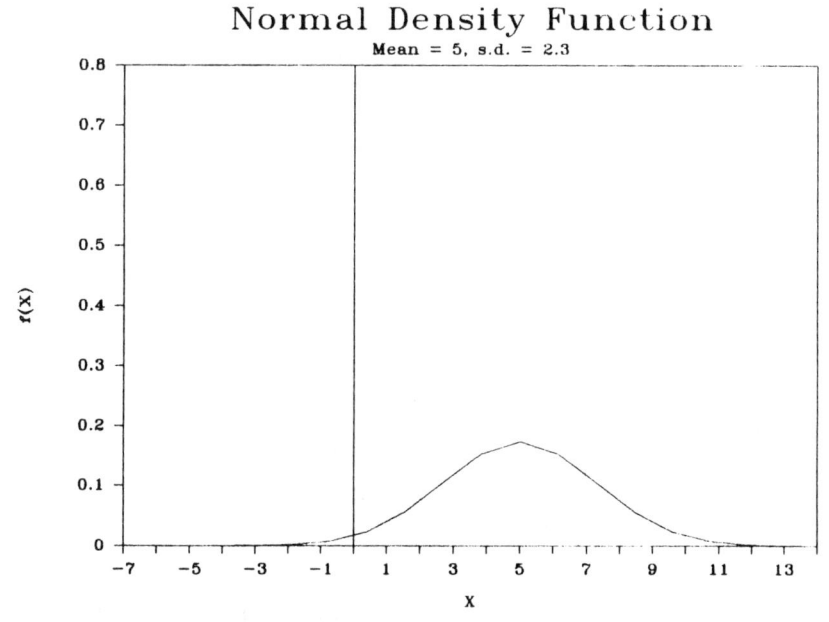

Fig. 8.5 Normal Density Function, mean=5, s.d.=2.3

The **normal** distribution is defined by the density function

$$f(X) = \frac{1}{\sigma\sqrt{(2\pi)}} e^{(X-\mu)^2 / 2\sigma^2} \qquad [8.6]$$

where **X** is a continuous variable taking any value between minus and plus infinity, and where also μ is the mean of the theoretical distribution and σ

its standard deviation. The symbols π and **e** are mathematical constants. They are obtained in 1-2-3 from @ functions.

Try entering **@PI** in an empty cell of your spreadsheet. The value 3.141592 will be displayed. Since **e** is often used with an exponent, as it is in equation 8.6, the 1-2-3 function **@EXP** requires an exponent as an argument. Try entering @EXP(1) in another cell of your spreadsheet. The value of 2.718281 will be displayed, which is the value of e^1 or **e**.

Table 8.4 Normal Density Function

	A	B	C	D	E	F
1	Normal	Density	Function			
2						
3			s.d.^2	s.d.*@SQRT(2*@PI)		1/const
4	Mean	s.d.	sigsq		const	k
5	5	2.3	5.29		5.765245	0.173453
6						
7	Mean-	Cum.	A+B,	k*@EXP(-1/(2*sigsq)*(X-Mean)^2)		
8	4*s.d.	s.d./2	X	f(X)		
9						
10	-4.2	0.0	-4.2	0.000058		
11	-4.2	1.2	-3.0	0.000379		
12	-4.2	2.3	-1.9	0.001927		
13	-4.2	3.5	-0.7	0.007621		
14	-4.2	4.6	0.4	0.023474		
15	-4.2	5.8	1.6	0.056312		
16	-4.2	6.9	2.7	0.105205		
17	-4.2	8.1	3.9	0.153072		
18	-4.2	9.2	5.0	0.173453		
19	-4.2	10.4	6.2	0.153072		
20	-4.2	11.5	7.3	0.105205		
21	-4.2	12.7	8.5	0.056312		
22	-4.2	13.8	9.6	0.023474		
23	-4.2	15.0	10.8	0.007621		
24	-4.2	16.1	11.9	0.001927		
25	-4.2	17.3	13.1	0.000379		
26	-4.2	18.4	14.2	0.000058		

We shall use 1-2-3 to graph the density function, and will observe how the graph changes as μ and σ are changed. Table 8.4 shows a suitable spreadsheet for plotting the graph. Its last two columns of figures contain values of **X** and of f(**X**) which we shall plot against one another. Let us see how this spreadsheet has been constructed.

We have defined **X** to take values between minus and plus infinity, but in fact **X** is almost certain to lie within 4σ of the distribution mean, μ, so values in this range have been set up. To do this, Range Name two cells to hold the **Mean**, μ, and **s.d.**, σ, and enter their values, shown as 5 and 2.3 respectively. Now set up a column of values of **+Mean**-4***s.d.** Each cell will then contain the lowest **X** value which we want to graph. To obtain **X** values in steps of **s.d.**/2, form a column containing the various amounts to be added on to the lowest **X** value. For the first row this is 0; for subsequent rows it is the value in the row above +**s.d.**/2 . Summing the first two columns gives the **X** values shown.

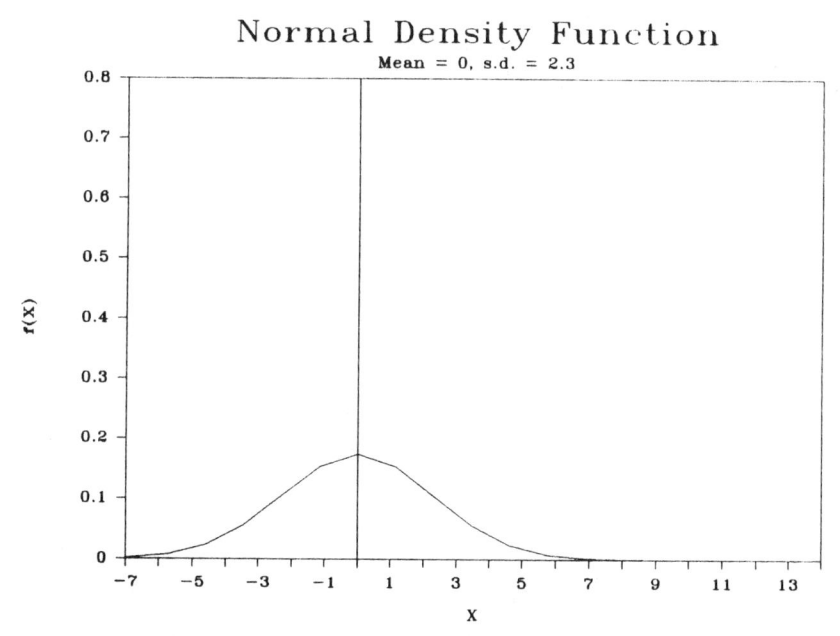

Fig. 8.6 Normal Density Function, mean=0, s.d.=2.3

The f(**X**) column contains the values of the density function defined in equation 8.6 corresponding to each of our **X** values. The constants required for the function have been evaluated separately in table 8.4, where cells Range Named **sigsq, const** and **k** hold values of σ^2, $\sigma\sqrt{(2\pi)}$ and $1/(\sigma\sqrt{(2\pi)})$ respectively. The formulae used to calculate these are: **s.d.**^2 for **sigsq**; **s.d.***@SQRT(2*@PI) for **const**; and 1/**const** for **k**.

The column of f(**X**) values has been calculated by entering and copying the formula

   +**k***@EXP(-1/(2***sigsq**)*(**X-Mean**)^2)

Fig 8.5 can be graphed from table 8.4 as an XY graph with the **X** values constituting the X range and the f(**X**) values the A range. Notice that Options Scale Y Scale Manual has been used with Lower Limit 0 and Upper Limit 0.8 to set the scale on the Y axis. After defining these limits select Quit, then Scale X Scale Manual to set the X axis to have Lower Limit -7 and Upper Limit 14. Quit from the Scaling and Options sub menus, and your graph should have the form shown in fig 8.5.

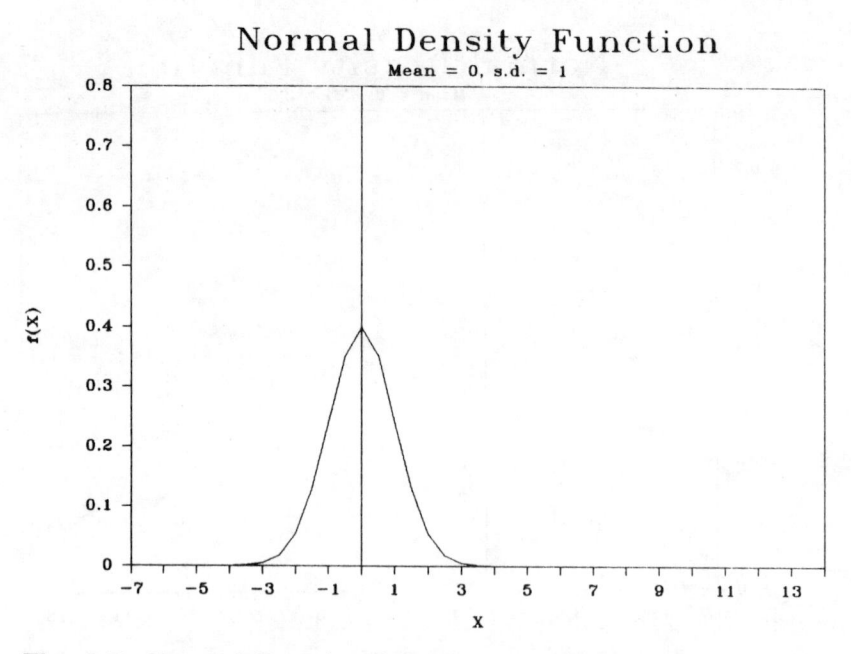

Fig. 8.7 Normal Density Function, mean=0, s.d.=1

Quit from the Graph menu and re-enter the Mean of the distribution as 0. Return to Graph and View your graph again; as in fig 8.6 it should now be centred at 0. Again Quit from the Graph menu and this time alter the s.d. to 1. The resulting graph is shown in fig 8.7. Notice that, had automatic scaling been used, all three graphs would have appeared the same, although the scales marked on the axes would have been different.

The change to a Mean of 0 and s.d. of 1 is the transformation usually made to standardise a normal distribution so that probabilities can be found by consulting tables. An alternative method of obtaining normal probabilities is to set up and use the spreadsheet described in the following section.

8.4 Continuous Distributions Probabilities

In later chapters various statistical tests and distributions will be used. This section shows how to find the associated significance probablities which tell us the confidence we can place in the results we have derived.

Throughout this book we have stressed how spreadsheets allow us to build up results step-by-step from first principles. In particular we have used definitional, rather than computational, formulae. However, when it comes to significance probabilities using continuous distributions this approach encounters a problem, since the relevant formulae involve integration (the measurement of areas under curves). This is signified by the elongated S integral symbol, which means add up an infinite number of infinitesimally small areas. Integration is not directly possible with a computer so we must use an approximation.

There are a variety of different approximations that can be used to generate significance probabilities. We shall use one developed by Landram, Cook & Johnston (1986) which gives answers correct to between 2 and 4 decimal places. There are more accurate ones available but they take the computer longer to evaluate and you longer to type in!

Enter table 8.5 into a worksheet. We shall use it to obtain probabilities for four different distributions, beginning with that for **F**.

Table 8.5 Probability Calculation Labels

	A	B	C
1	PROBABILITY CALCULATIONS		
2	Label:	Value:	
3	C	1	Code: F=1, t=2, z=3, Chisq=4
4	N	2	Numerator DF, blank if z or t
5	D	3	Denominator DF, blank if z or Chisq
6	S	9.55	Statistic value
7	E		
8	F		
9	T		Transformed statistic
10	G		
11	H		
12	J		
13	A		
14	B		
15	K		
16	L		
17	Z		
18	P		Probability
19	Q		

Use Range Name Labels Right on the letters in column A to name the individual cells in column B. In cells B7 to B19 enter the formulae in table 8.6.

Table 8.6 Probability Calculation Formulae

```
Cell: Formula:
B7:   @CHOOSE(C-1,N,1,1,N)
B8:   @CHOOSE(C-1,D,D,1000,1000)
B9:   @CHOOSE(C-1,S,S*S,S*S,S/E)
B10:  @IF(T<1,F,E)
B11:  @IF(T<1,E,F)
B12:  @IF(T<1,1/T,T)
B13:  2/(9*G)
B14:  2/(9*H)
B15:  @ABS(((1-B)*(J^(1/3)))-(1-A))
B16:  +K/@SQRT(B*(J^(2/3))+A)
B17:  @IF(C=3,S,@IF(H<4,L*(1+(0.08*(L^4)/(H^3))),L))
B18:  0.5/(((1+(Z*0.196854)+(Z^2*0.115194)+(Z^3*0.000344)
      +(Z^4*0.019527))^4)
B19:  +P/2
```

After adding labels you should end up with a spreadsheet like table 8.7. Notice that column B shows results to 10 decimal places in order for you to check that you have got the formulae exactly right. You can then display just 2 or 3 decimal places for general usage.

Table 8.7 Probability Calculations

```
         A            B           C
1    PROBABILITY CALCULATIONS
2    Label:        Value:
3         C             1  Code: F=1, t=2, z=3, Chisq=4
4         N             2  Numerator   DF, blank if z or t
5         D             3  Denominator DF, blank if z or Chisq
6         S          9.55  Statistic value
7         E             2
8         F             3
9         T          9.55  Transformed statistic
10        G             2
11        H             3
12        J          9.55
13        A  0.1111111111
14        B  0.0740740741
15        K  1.0755749326
16        L  1.6131909582
17        Z  1.6455618839
18        P  0.0497418194  Probability
19        Q  0.0248709097
```

This shows that an **F** value of 9.55 with a numerator degrees of freedom of 2 and a denominator degrees of freedom of 3 has a significance probability of approximately 0.05.

8.5 t Values

To find the signficance probability of a **t** value, enter the value 2 into B3. Range Erase B4 and enter the relevant degress of freedom into B5. Enter the **t** value itself into B6 and the two-tailed probability of a value greater than or equal to it occurring will appear in B18. The one-tailed result will appear in B19.

8.6 Z Values

To find the signficance probability of a **Z** value, enter the value 3 into B3. Range Erase B4..B5 and enter the **Z** value itself into B6, when the resulting probability will appear in B18. Note that this is the area in the right hand tail of the distribution.

8.7 Chi-squared Values

To find the signficance probability of a **Chi-squared** value, enter the value 4 into B3 and the relevant degrees of freedom into B4. Range Erase B5 and enter the Chi-squared value itself into B6, and the resulting probability will appear in B18. Note that this is the area in the right hand tail of the distribution.

8.8 Summary

Although it is unfortunate that we must use an approximation in order to generate significance probabilities, this approach does serve to stress that there are fundamental relationships between the **F, t, Z** (normal) and Chi-squared distributions, which many elementary statistical textbooks do not discuss. It is also unfortunate that different textbooks present statistical tables in different ways. In particular, they often require manual interpolation between values, whilst the approach given here will automatically give an actual significance probability.

In this chapter we have studied various probability distributions. Using the graphical and recalculation facilities of 1-2-3 we have been able to see immediately the effects of altering the values which specify a distribution. To obtain meaningful graphs showing the standardisation of a normal variable we used the Graph Options Scale command.

We have used @VLOOKUP to look up values in a table, and @PI and @EXP to obtain the appropriate mathematical constants.

9
Decision Analysis

9.1 Introduction

Decision analysis sets out rational rules for deciding between alternative **actions** when the **payoffs** they will yield are uncertain. The uncertainty arises because each action may result in a different payoff depending on which of various possible **outcomes** or **states of the world** actually happens to occur. We know that one will come about, but we do not know, and cannot influence, which it will be. We treat these outcomes, therefore, as chance events. The payoffs which would arise for each outcome are listed in a **payoff** or **profit table**. Various possible decision criteria can be used for choosing which action to take. We shall consider only the **Bayesian** criterion of **maximising expected payoff** or **profit**.

Decision analysis augments and clarifies, but the quality of the decision made naturally depends on the quality of the information on which it is based. Skill and judgement are required in obtaining appropriate and accurate data and in incorporating relevant psychological factors.

9.2 Expected Payoff

The example below shows how a payoff table is drawn up and the decision analysis carried out.

Consider a 20 year old woman with her first car, deciding whether she should take out third party insurance, or whether she should pay an additional premium of £100 for comprehensive cover. She is choosing, then, between two possible actions, the different levels of insurance cover. She considers three possible events that might occur in the first year, and how much each would cost her in addition to the cost of insurance. If she were to be responsible for a write-off crash, it would cost £5000 to replace her car; repairs following a minor crash might cost £500. If she chose third party insurance she would have to pay these amounts, while if she chooses comprehensive insurance she would pay nothing. If she were to have no accident she would pay nothing for accident repairs, regardless of the type of insurance she chooses. Her annual income less the cost of third party insurance is £10,000. From the above information she draws up the payoff table shown in table 9.1, which lists her income net of insurance and accident repair costs for each possible action/outcome combination.

Table 9.1 Payoff Table (£'000 per annum)

```
            A              B           C           D
1   Car insurance decision problem
2   ------------------------------
3
4   Payoff table (£'000) per annum
5   ------------------------------
6                          Insurance chosen
7                          Comp        3rd party
8   Write-off crash         9.9         5.0
9   Minor crash             9.9         9.5
10  No crash                9.9        10.0
```

We see from the table that if she has no crash, third party only insurance would give her the higher payoff, but if she has a crash she would do better if she has chosen comprehensive cover.

Table 9.2 Payoff Table and Subjective Probabilities

```
            A              B           C           D           E           F
1   Car insurance decision problem
2   ------------------------------
3
4   Payoff table (£'000) per annum
5   ------------------------------
6                          Insurance chosen        Subjective prob
7                          Comp        3rd party   specified event
8   Write-off crash         9.9         5.0         0.01
9   Minor crash             9.9         9.5         0.09
10  No crash                9.9        10.0         0.90
11                                                 ---------
12                                                  1.00
```

If the motorist is to choose which insurance cover she should take out according to the Bayesian criterion, she must estimate the probabilities of her having a write-off crash, a minor crash, and no crash. She will use the subjective approach in estimating these, that is, they will be her personal assessments of the likelihoods of the various outcomes. Let us suppose that the values she arrives at are those shown in column E of table 9.2.

She then calculates separately the expected payoffs for each of the choices open to her. These are displayed in C13 and D13 of table 9.3. To find the expected payoff for comprehensive insurance she must multiply each element in the column headed **comp** by its probability, shown in the **subjective prob** column, and add these results. The expected payoff in this instance is of course 9.9 (in units of £'000), since this value is certain to occur whether or not she has a crash. We shall, however, enter a general formula to calculate the expected value, so that if she were to alter the payoffs to take into account the effect of a crash on her no-claims discount, the expected payoff would be recalculated correctly.

Table 9.3 Expected Payoffs Calculated

	A	B	C	D	E	F
1	Car insurance decision problem					
2	------------------------------					
3						
4	Payoff table (£'000) per annum					
5	------------------------------					
6			Insurance chosen		Subjective prob	
7			Comp	3rd party	specified event	
8	Write-off crash		9.9	5.0	0.01	
9	Minor crash		9.9	9.5	0.09	
10	No crash		9.9	10.0	0.90	
11					---------	
12					1.00	
13	Expected payoff		9.9	9.905		

The formula entered in C13 is

+C8*$E8+C9*$E9+C10*$E10

Notice

(i) the + sign at the beginning to indicate to 1-2-3 that a value is being entered.

(ii) the use of relative and mixed absolute/relative cell addresses to enable the formula to be copied.

(iii) these addresses can be obtained by pointing in the relative form. To convert them to be wholly or partly absolute the [4/ABS] function key is depressed until the desired form is shown.

(iv) with this formula we shall be able to alter any of the values we have entered and the expected payoff will immediately be recalculated correctly.

The larger of the expected payoffs shown in table 9.3. is the value for third party insurance, 9.905 or £9,905. According, therefore, to the Bayesian criterion of maximising expected payoff, the motorist should choose third party only insurance.

Bayesian analysis is often criticised for the inaccuracy of the subjective probabilities used. You might like to try some "What if?" calculations using different subjective probabilities, each set summing to 1, to see their effects on the expected payoffs and thus on the motorist's decision.

9.3 Posterior Expected Payoff

Suppose that before the motorist makes her decision someone tells her that drivers aged under 21 are more likely than others to have crashes. She then feels that she should modify her subjective probability estimates to take into account what information she can get about the effect of her age, and she may use Bayes theorem to do this.

The information she incorporates will take the form of **conditional probabilities** or **likelihoods**. These can usually be measured more objectively than can the original subjective probabilities of the events. The conditional probabilities may be obtainable from past records. Alternatively, when items are sampled these probabilities have to be calculated using an appropriate distribution such as the binomial, for which you should use the method given in Chapter 8.

Let a write off crash, a minor crash and no crash be denoted events B_1, B_2 and B_3 respectively, and her being aged under 21 be denoted as event A. Modifying formula 7.12 to take into account that there are three events, B_1, B_2, B_3, which could occur we can calculate the posterior probability of her having a write-off crash given that she is aged under 21, $P(B_1/A)$, as follows:

$$P(B_1/A) = \frac{P(B_1)P(A/B_1)}{P(B_1)P(A/B_1)+P(B_2)P(A/B_2)+P(B_3)P(A/B_3)} \quad [9.1]$$

Notice that the numerator of this expression also forms the first term of the denominator. We shall require also the posterior probabilities of a minor crash, $P(B_2/A)$, and of no crash, $P(B_3/A)$. These can be similarly calculated with the same denominator as equation 9.1, their numerators being respectively the second and third terms of that denominator.

These calculations may appear very complicated, but in fact the three posterior probabilities can be found simultaneously and the working can be set out in a concise table. We have already entered in the spreadsheet the values of $P(B_1)$, $P(B_2)$ and $P(B_3)$. These are the subjective probabilities of the events, sometimes called prior probabilities, which we are going to modify in the light of the information about the motorist's age. To do this according to formula 9.1, we require the conditional probabilities $P(A/B_1)$, $P(A/B_2)$ and $P(A/B_3)$. These are the probabilities that someone who has been responsible for a write-off crash, B_1, is aged under 21, that a person who has had a minor crash is in that age group and also that someone who has had no crash is in that age group. These values might be calculable from insurance company records. We shall assume that they are as set out in column G of table 9.4 which shows that when a write-off crash, B_1, has occurred the probability that it was caused by someone aged under 21 is 0.25.

Table 9.4 Calculation of Posterior Probabilities

```
              E           F           G          H            I
 6    Subjective prob   Prob u21    Joint    Posterior prob
 7    specified event   /event    prob, ExG    H/sum(H)
 8         0.01           0.25      0.0025      0.0371
 9         0.09           0.12      0.0108      0.1605
10         0.90           0.06      0.0540      0.8024
11    ---------                    ---------
12         1.00                      0.0673
```

We now require to find $P(B_1).P(A/B_1)$ which, according to the Multiplication Rule, is the joint probability of B_1 and A both occurring. To do this we enter a formula in H8 which will multiply together the contents of cells E8 and G8. This formula may then be copied to H9 and H10 to find $P(B_2).P(A/B_2)$ and $P(B_3).P(A/B_3)$ respectively. The denominator of formula 9.1 is, therefore, the sum of the values in H8, H9 and H10, and a formula is entered in H12 to calculate it.

We can now calculate the posterior probabilities which are displayed in column I of table 9.4. To obtain $P(B_1/A)$ we enter a formula in I8 to divide the value in H8 by the total of the joint probabilities held in H12. We would like to copy this formula to I9 and I10 to find the other posterior probabilities and so must make absolute the row of the cell (I12) containing the divisor.

The posterior probabilities which are thus formed in column I are the revised probabilities of the motorist being responsible for a write-off crash, a minor crash and no crash, respectively. We notice that, because write-off crashes were more likely to have involved a driver aged under 21, the posterior probability of this event is higher than was the prior, subjective, one.

We can now calculate the posterior expected payoffs of choosing each type of insurance cover, using the same method of calculating expectation as before but substituting the posterior probabilities for the subjective ones. A formula to do this is entered in C16 and copied to D16, as shown in table 9.5. Since with comprehensive insurance the payoff of £9,900 is certain, we find that this value is both the expected payoff calculated earlier and also the posterior expected payoff. For third party insurance, however, the posterior expected payoff at £9,734 is lower than the expected payoff calculated earlier of £9,905, and because it is now also lower than the payoff from comprehensive insurance the best action now appears to be to choose the comprehensive policy. That is, using the Bayesian criterion of maximising posterior expected payoff the motorist should choose comprehensive insurance.

Table 9.5 Posterior Expected Payoffs Calculated

```
         A          B              C          D            I
 1  Car insurance decision problem
 2  ------------------------------
 3
 4  Payoff table (£'000) per annum
 5  ------------------------------
 6                          Insurance chosen      Posterior prob
 7                          Comp      3rd party     H/sum(H)
 8  Write-off crash         9.9         5.0         0.0371
 9  Minor crash             9.9         9.5         0.1605
10  No crash                9.9        10.0         0.8024
11
12
13  Expected payoff         9.9         9.905
14
15  Posterior
16  Expected payoff         9.9         9.734
```

We notice how the changes that have taken place accord with commonsense. The information which became available indicated that the more serious crashes were more likely to be caused by drivers aged under 21. This caused the motorist to increase her estimate of the probability of her having such a crash, and this led to the comprehensive insurance then appearing preferable to the third party insurance which had previously seemed the better choice.

9.4 Summary

Decision analysis is a useful approach for evaluating the gains or losses from alternative courses of action. Since both the payoff figures and the subjective probabilities used are liable to error, the spreadsheet's recalculation facilities are particularly useful. You can easily carry out a sensitivity analysis to see the effects of various changes on the optimal action.

10
Sampling and Inference

- **Move command**
- **@RAND**
- **Manual recalculation**

10.1 Introduction

In **statistical inference** we define a **population**, which is the complete collection of items in which we are interested, e.g. all the people in a country, all the records in a database, or all the items produced by a particular machine in a certain time period. We select a **simple random sample** of items from our population, that is, the items are chosen in such a way that every item in the population has an equal chance of being included in the sample. On the basis of our sample values we can then make probabilistic statements about the population.

The methods used can be justified by statistical theory, but the concepts involve visualising a repetition of the sampling process, which is difficult to imagine. With a spreadsheet we can actually perform such a repetition, and investigate what happens. We can look to see whether the theoretical results hold good in a particular case, and in doing so gain a greater understanding of their meaning.

In this chapter we shall enlarge our earlier database to act as a population. We shall be concerned with the mean and standard deviation of the salary values, and can, of course, compute these from the population data. We shall then see how to use 1-2-3 to generate random numbers for selecting a random sample of salaries from our population, and we shall compute the

Table 10.1 Extended Database

	A	B	C	D	E	F
1	EMPLOYEE PERSONAL DATA					
2	--					
3	ID:	SURNAME:	FORENAME:	SALARY:	SEX:	TIME:
4	1	Smith	Angela	4800	F	PART
5	2	Jones	Roberta	5110	F	PART
6	3	Reed	Jane	5520	F	PART
7	4	Mackie	Charles	5570	M	PART
8	5	Soper	Jean	6325	F	PART
9	6	Archer	Sean	6750	M	PART
10	7	Phillips	John	6785	M	PART
11	8	Evans	Paula	7320	F	PART
12	9	Lawson	Hugh	7320	M	PART
13	10	Smith	Julia	7320	F	PART
14	11	Thatcher	Edward	8894	M	PART
15	12	Lee	Martin	10655	M	FULL
16	13	Young	Peter	10760	M	FULL
17	14	White	Lee	12480	M	FULL
18	15	Cox	Henry	13560	M	FULL
19	16	Baker	Rachel	14380	F	FULL
20	17	Bingham	John	14380	M	FULL
21	18	Brown	John	15460	M	FULL
22	19	McKay	Iain	20860	M	FULL
23	20	Jeffreys	Sue	9500	F	PART
24	21	Quarmby	Rhoda	10500	F	FULL
25	22	Macklin	Jackie	11500	F	FULL
26	23	Ralhan	Veena	12500	F	FULL
27	24	Palin	Petrina	13500	F	FULL
28	25	Andrews	Derek	14500	M	FULL
29	26	Peacock	Derek	15500	M	FULL
30	27	Andrews	Angela	4800	F	PART
31	28	Archer	Roberta	5025	F	PART
32	29	Baker	Jane	5690	F	PART
33	30	Bingham	Charles	5315	M	PART
34	31	Brown	Jean	6665	F	PART
35	32	Cox	Sean	6325	M	PART
36	33	Evans	John	7295	M	PART
37	34	Jeffreys	Paula	6725	F	PART
38	35	Jones	Hugh	8000	M	PART
39	36	Lawson	Julia	6555	F	PART
40	37	Lee	Edward	9744	M	PART
41	38	Mackie	Martin	9720	M	FULL
42	39	Macklin	Peter	11780	M	FULL
43	40	McKay	Lee	11375	M	FULL
44	41	Palin	Henry	14750	M	FULL
45	42	Peacock	Rachel	13105	F	FULL
46	43	Phillips	John	15740	M	FULL
47	44	Quarmby	John	14015	M	FULL
48	45	Ralhan	Iain	22390	M	FULL
49	46	Reed	Sue	7885	F	PART
50	47	Smith	Rhoda	12200	F	FULL
51	48	Smith	Jackie	9715	F	FULL
52	49	Soper	Veena	14370	F	FULL
53	50	Thatcher	Petrina	11545	F	FULL
54	51	White	Derek	16540	M	FULL
55	52	Young	Derek	13375	M	FULL

sample mean and standard deviation to compare with the population values. We shall use our sample values to calculate an interval estimate for the population mean.

Table 10.2 Extended Database Construction Columns

	G	H
1		
2	---------	---------
3	STEP 85	SAL+OR-
4	0	4800
5	85	5025
6	170	5690
7	255	5315
8	340	6665
9	425	6325
10	510	7295
11	595	6725
12	680	8000
13	765	6555
14	850	9744
15	935	9720
16	1020	11780
17	1105	11375
18	1190	14750
19	1275	13105
20	1360	15740
21	1445	14015
22	1530	22390
23	1615	7885
24	1700	12200
25	1785	9715
26	1870	14370
27	1955	11545
28	2040	16540
29	2125	13375

Having set up these calculations for one sample we shall then use 1-2-3's recalculation facility for obtaining further random samples and making the same calculations for each. We shall build up a column of sample mean values, and compare their distribution with the distribution of the salary values in the population. We shall look to see whether the theoretical relationship between these distributions is apparent in our results.

10.2 Extending our Database

- **Move command**

The database which we set up in table 7.1 contains 26 records. This is rather small to use as a population, and we shall extend it to the 52 items shown in table 10.1. We will then be able to make inferences from a sample of 5 records without using the finite population correction. (The finite population correction is required when the sample size exceeds 10% of the population).

Notice that, whereas in table 7.1 the ID: column was at the right of the database, in table 10.1 it forms the leftmost column. This is necessary if we are to use the @VLOOKUP function. Table 10.2 contains two columns of intermediate calculations, STEP 85 and SAL+OR-, which you can use to derive the additional rows of the database without a lot of extra typing.

When you have retrieved the database shown in table 7.1, move the pointer to column A and use the Worksheet Insert command to insert a blank column there. Using the **Move** command, move the title and its underlining one cell to the left, and move the ID: column into the empty left hand column. When moving a range, as when copying one, it is necessary only to define the top left hand corner of the TO Range and entries will be made in, below and to the right of it, forming the same size of range as the FROM range.

Notice that before using the Move command we created a blank space to receive the entries we wished to move. This was necessary because the Move command overwrites any previous entries in the Move To range. The Move command is useful and powerful, but can be destructive. Always ensure that you have Saved the worksheet before using it.

Now Copy the 26 records to form a second set immediately below the first, and use Data Fill to obtain ID:'s 27 to 52. Sort the lower set of surnames into alphabetical order, defining the Data-Range as just the surnames you want to be sorted, and the Primary-Key address as any cell in that column.

Now form the new salary values by first using Data Fill with Start 0, Step 85 to create a STEP 85 column, filling in values to correspond to each of the original 26 records. The new salary values are formed in the column headed SAL+OR- by alternately adding and subtracting the STEP 85 values from successive SALARY: values. Enter formulae to add the values in the first data row and to subtract STEP 85 from SALARY: in the second. Copy these formulae to the next two rows, then these four formulae to the next four rows, and so on. Now replicate the SAL+OR- values in the SALARY: column from record 27 downwards by entering a formula in cell D30 consisting of +H4, the address of the first SAL+OR- value, and Copying it down the SALARY column. Complete your spreadsheet as in table 10.1 by copying to the right the line above the column headings. Range Name the extended database in the first six columns **DBE**.

10.3 Population Summary Values

We shall regard the 52 values listed in the SALARY: column of table 10.1 as the **population** of **X** values, and we can Range Name them **X**.

Since we know what all the populaion values are, we can, of course, calculate the population mean using @AVG and the standard deviation using @STD. These are calculated at the foot of the salary column and shown in table 10.3. As the population standard deviation is often denoted σ, we shall Range Name it **Sigma**.

For illustrative purposes we are going to choose random samples from this population, each consisting of 5 values. We shall estimate the population mean and standard deviation from each sample, and compare these estimates with the exact values of £10,430.63 and £4,141.17 respectively.

10.4 Simple Random Sampling

- @RAND
- Worksheet Global Recalculation Manual
- Function key [9/CALC]

We shall choose a simple random sample of five salary values by generating five random numbers and using the @VLOOKUP function to select the salary values from the records corresponding to each.

The **@RAND** function in 1-2-3 generates a random number between 0 and 1. Because decimal numbers are generated we require random numbers between 1 and 53, the integer parts of which, being effectively between 1 and 52, will identify the records we require. We shall use the formula:

    52*@RAND+1   [10.1]

Multiplying @RAND by 52 generates a random number between 0 and 52, so that adding 1 then gives a number in the desired range. Enter a label to identify a row as containing your random numbers, then enter formula 10.1 in an adjacent cell. Watch what happens to the value in this cell as you copy the formula four cells to the right to obtain a total of five random numbers, each of which is between 1 and 53, such as are shown in table 10.3. It is, of course, very unlikely that you will happen to obtain exactly the same set of random numbers.

You should have noticed that as you activated the Copy command the first random number changed. To ensure that the random numbers change only when you want them to do so, you must set 1-2-3's recalculation facility to manual. To do this, select from the menu **Worksheet Global Recalculation Manual**. Now the worksheet will generate new random numbers and recalculate as a consequence of altered values only if the **[9/CALC]** key is pressed. As you proceed with building up the worksheet the CALC indicator will come on in the bottom right of your screen. Don't worry about this. It indicates to you that a value has changed and you may wish to press [9/CALC] to see its effect. In our case, however, it will come on because the random numbers have changed, and we shall only actually want new random numbers when we choose another sample.

Table 10.3 Random Sample and Summary Statistics

```
         B           C          D          E          F          G
57   @AVG(X)                 10430.63                 5
58   @STD(X)                  4141.17                 n
59                            Sigma              1851.99
60                                              Sigma/@SQRT(n)
61   Rand No:
62   14.80858  9.317424  16.60114  25.07779  3.148876
63   Samp X's:
64      12480      7320     14380     14500      5520   10840.00
65                                                         Xbar
66            @SQRT(@STD(Samp X's)^2*n/(n-1))           4162.62
67                                                            s
68                                    s/@SQRT(n)       1861.58
69                                                      se(Xbar)
70   2.776
71   t         Confidence limits:Xbar-t*se(Xbar)       5672.25
72                               Xbar+t*se(Xbar)      16007.75
```

The first random number shown in table 10.3 is 14.80858. We can use @VLOOKUP to find the salary in table 10.1 which corresponds to an ID: of the integer part of this number, that is, an ID: of 14. Referring to table 10.1 we can see that this value is £12,480.

If the first random number is in B62, and if the extended database has been named **DBE**, we enter

@VLOOKUP(B62,$DBE,3) [10.2]

The lookup range is the extended database, **DBE**. The function locates the first number in the first column of the range, ID:, which is larger than the random number contained in cell B62. It then backs up to the previous row. With B62 containing the value 14.80858, then, 1-2-3 locates 15 in the ID: column, and backs up to 14. Thus it automatically selects the row corresponding to the integer part of the random number.

The third argument of the @VLOOKUP function gives the number of the column, SALARY:, whose value will be selected. Numbering from 0 from the left it has the value 3. Hence the 1-2-3 formula given in 10.2 will select from the values shown in table 10.1 the salary £12,480 corresponding to an ID: of 14, which is the integer part of the random number 14.80858. Copying the formula four cells to the right selects a total of 5 salaries, labelled and named **Samp X's** in table 10.3.

10.5 Confidence Limits

We shall now calculate the sample mean and standard deviation, which may be used as estimates of the population values when the latter are unknown. In this instance we have already found the population values, so we shall be able to compare the sample estimates with them. Using these sample estimates we shall set up 95% confidence limits for the population mean, and check whether the population mean lies within them.

It is usual to write **n** for the sample size. Applying @COUNT to the **Samp X's** gives a value of 5, and this count is named **n** in table 10.3.

The mean of the sample values is calculated in the usual way using the @AVG function. Its value, 10840.00, is displayed in table 10.3 to the right of the sample values, and it is named **Xbar**.

The inbuilt @STD function, however, needs adjustment for calculating the sample standard deviation. We noted in section 4.5 that this function uses **n** as divisor, but that when estimating a population standard deviation from sample values we require a divisor of **n**-1. In other words, the sample standard deviation, **s**, is defined as:

$$s = \sqrt{\left[\sum_{i=1}^{n} (X_i - Xbar)^2 / (n-1)\right]} \qquad [10.3]$$

To obtain **s** from 1-2-3's @STD function we must square the value calculated by the latter, adjust the divisor by multiplying by **n/(n-1)** and take the square root of the result. The 1-2-3 formula used for **s** in table 10.3 is:

@SQRT(@STD(**Samp X's**)^2*n/(n-1))

From the sample values shown in table 10.3 this gives a value of 4162.62 for the standard deviation estimate, which we name **s**. We then estimate the standard error of the mean, named **se(Xbar)**, using +s/@SQRT(n).

We can now use the measures calculated from our sample to set confidence limits for the population mean. For this, since our sample is small, we must use the **t** distribution with **n-1** degrees of freedom. For 4 degrees of freedom the value of the **t** distribution which cuts off an area of 0.025 to its right, and so yields (considering both tails of the distribution) 95% confidence limits, is 2.776. This value is entered in table 10.3 and named **t**. The lower limit for the population mean is then found using the formula +**Xbar-t*se(Xbar)**, and the upper one using +**Xbar+t*se(Xbar)**. The values shown in table 10.3 give a 95% confidence interval for the mean salary of all persons in the population of £5,672.25 to £16,007.75.

Let us now compare the values we have calculated from our sample of 5 with the true population values. There is some difference between the sample mean of £10,840 and the population one of £10,430.63. The sample standard deviation of £4,162.62 rather over-estimates that of the population, which is £4141.17. Essentially, because our sample size is so small, the precision of our estimates is not very great. A further consequence of this is that the confidence intervals we set are very wide, but you should notice that those shown in table 10.3 do in fact include the population mean.

10.6 Sample Replication

Pressing the [9/CALC] key will give you a new set of random values, the corresponding new sample values, and recalculated values for all the values calculated from them. You can compare the values calculated from your new sample with those from your previous sample, and with the population values, and you can then press the [9/CALC] key again for another sample. As you repeat the sampling process, check to see how many of your sets of confidence limits include the population mean. Since they are 95% limits, the theoretical expectation is that out of every 100 sets of such limits that you construct, 95 of them will contain the population mean.

As you continue drawing random samples of the salary values, build up a column containing the various sample mean values that you obtain. Table 10.4 shows a column containing 20 successive **Xbar** values. As you bring each **Xbar** value into the column, convert it from a formula to a numerical value by locating the pointer on it and pressing [2/EDIT] followed by [9/CALC] and then [RETURN].

This prevents it from recalculating as you obtain your next sample.

Table 10.4 Sample Means and their Distribution

```
              C          D          E          F           G          H
57                   10430.63                  5
58                    4141.17                  n
59                    Sigma               1851.99
60                                      Sigma/@SQRT(n)
61
62    9.317424   16.60114   25.07779   3.148876              Successive
63                                                              Xbars
64         7320       14380      14500       5520   10840.00   14016.0
65                                                      Xbar   13076.0
66    @SQRT(@STD(Samp X's)^2*n/(n-1))              4162.62     10910.0
67                                                         s   12314.0
68                             s/@SQRT(n)          1861.58      8043.0
69                                               se(Xbar)     12112.0
70                                                             6856.0
71    Confidence limits:Xbar-t*se(Xbar)            5672.25     10646.0
72                       Xbar+t*se(Xbar)          16007.75      9868.0
73                                                            10068.8
74                                                            11742.0
75                                                             8286.0
76                                                            11896.0
77                                                             9301.0
78                                                             8905.0
79                                                             7248.8
80                                                            11770.0
81                                                             9370.0
82    @AVG(Xbars)             10310.52                         8181.0
83    @STD(Xbars)              1958.11                        11600.0
```

An easy way of building up the column of **Xbars** is to enter the formula +$**Xbar** at the top and Copy it down as far as you wish. This will set up a column of values all equal to the current sample mean. Convert the first of these to a numerical value by making it the active cell then pressing [2/EDIT] followed by [9/CALC] and the [RETURN] key. Now press [9/CALC] again. This gives you a new sample and all but the first of the **Xbars** will change to become the new sample mean. Repeat the process of converting one value in the column to a numerical value, then obtaining a new sample until you have a column of fixed numerical values.

10.7 Sample Means Distribution

The Central Limit Theorem tells us that, for large samples, regardless of the population distribution, the distribution of the sample means is approximately Normal. The theoretical mean of the sample means distribution is the population mean, and its theoretical standard deviation is the population standard deviation divided by the square root of the sample size. The latter is usually written as $\sigma/\sqrt{n}$, but in table 10.3 it is denoted **Sigma/@SQRT(n)**.

Apply the @AVG and @STD functions to your **Xbars**. From a relatively small number of sample replications you cannot expect that the average of the **Xbars** will be exactly the population mean of £10,430.63. In table 10.4 it is shown as £10,310.52, which is quite close. Similarly, the standard deviation of the **Xbars** will not be exactly the theoretical value of **Sigma/@SQRT(n)**, which is £1,851.99, but in table 10.4 it is displayed as £1,958.11.

Table 10.5 Relative Freq. Distns. of Population & Means

	A	B	C	D	E	F	G
85	Distribution of Salary, X			Distribution of Means of			
86	in Database			Salaries in Samples, Xbars			
87							
88	Upper Class		Class	Rel	Upper Class		Rel
89	Limit	f	Midpoint	freq	Limit	f	freq
90							
91	5000	2	4000	3.85%	5000	0	0.00%
92	7000	13	6000	25.00%	7000	1	5.00%
93	9000	7	8000	13.46%	9000	5	25.00%
94	11000	7	10000	13.46%	11000	6	30.00%
95	13000	7	12000	13.46%	13000	6	30.00%
96	15000	10	14000	19.23%	15000	2	10.00%
97	17000	4	16000	7.69%	17000	0	0.00%
98	19000	0	18000	0.00%	19000	0	0.00%
99	21000	1	20000	1.92%	21000	0	0.00%
100	23000	1	22000	1.92%	23000	0	0.00%
101	25000	0	24000	0.00%	25000	0	0.00%
102		0				20	

With 1-2-3 it is easy to obtain a graphical display of the distribution of the population values, **X**, and of the **Xbars**. Construct relative frequency distributions, as we did in chapter 5, and graph each of these against the class midpoints. Table 10.5 shows the relative frequency distributions for the values in tables 10.1 and 10.4, and fig. 10.1 shows the line graph plotted from it.

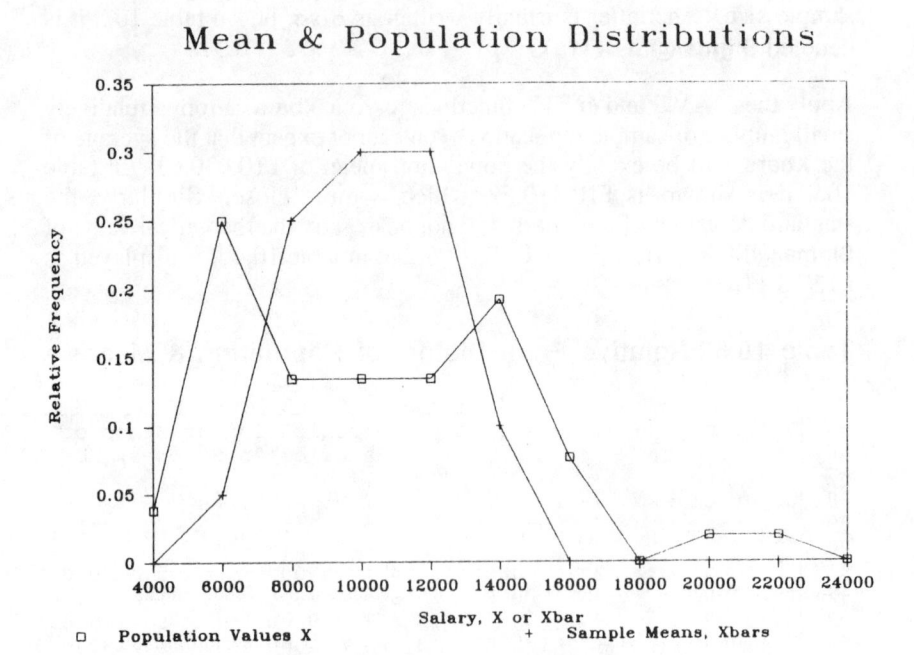

Fig. 10.1 Mean and Population Distributions

10.8 Summary

In this chapter we have learned a new command, Move, for altering the layout of a spreadsheet. It is powerful, so potentially very useful, but beware of its overwriting action, and always save a copy of your spreadsheet before using it.

We have found out how to generate random numbers within a desired range in 1-2-3, and how to look up a value in a column of a table. We have, for once, found ourselves not wanting 1-2-3's automatic recalculation facility, and have discovered how to change to manual recalculation.

These facilities have enabled us to simulate a repeated sampling process and we have been able to investigate the relationships which hold between population and sample values.

11
Crosstabulation of Data

- **Data Table**

11.1 Introduction

It is common in statistical analysis to present data in the form of frequency counts in various categories. In particular, sample survey data, often generated by questionnaires, will record such categorical items as sex, class and age group. The questionnaire responses are naturally stored in 1-2-3 in the form of a database where each record represents a case and each field a variable.

One way to look at such data is to form a table with one set of categories compared with another. This is known by various names: a **crosstabulation**, a **contingency table**, a **cross-classification** or a **two-way table**. These tables are characterised by having two or more column labels and two or more row labels arranged around four or more cells with frequency counts in each. These can be raw data or in the form of percentages of the total. Such tables can be generated using 1-2-3's **Data Table** command.

We have already encountered 1-2-3's Data Distribution option which gives a frequency distribution of a set of numerical values. However, that will only work for a complete set, rather than a specified sub-set, of input values. Furthermore, categorical responses would have to be coded numerically, e.g. for sex $M = 1$, $F = 2$. In this chapter we shall combine 1-2-3's database facility with a Data Table command to generate a crosstabulation by counting the numbers of persons in various category combinations. We shall then analyse whether our sample indicates that the row and column criteria are independent, using the **Chi-squared test**.

11.2 Database

We have already seen in Chapter 7 on probabilites how the Data Query command can be used to set up a database from which the @D functions can give summary information. So File Retrieve the worksheet which is shown in table 7.1. Copy the column labels from row 3 into row 31 to head a Criterion range. Range Name this row and the next as **CR**, for Criterion range, that is A31..F32. Notice in table 11.1 that row 32 is blank with no actual criteria present yet. Ensure that the database A3..F29 is named **DB**. In cell A35 enter the formula:

@DCOUNT($**DB**,2,$**CR**)

This means count those entries in the database as specified by the Criterion range. Notice that both the database and Criterion range are absolute names, whilst the offset, 2, refers to the salary column but could be any number from 0, the surname column, to 4, the time column. Because the Criterion range is empty, all items satisfy it and the total number of entries in the offset column, 26, is displayed in A35 when the formula is entered.

Table 11.1 Criterion Range and Database Count

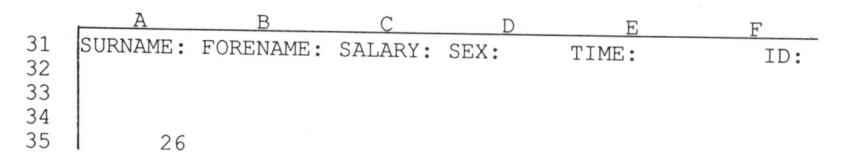

	A	B	C	D	E	F
31	SURNAME:	FORENAME:	SALARY:	SEX:	TIME:	ID:
32						
33						
34						
35		26				

11.3 Data Tables

- **Function key [8/TABLE]**

1-2-3's Data Table facility allows the construction of "What if?" tables, that is tables showing the results of substituting various values into formulae. This provides a simple way of doing sensitivity analyses and, in our case, crosstabulations.

1-2-3 has two types of Data Tables: 1 and 2. The first type tabulates the effect on various formulae of substituting a series of input values into a specified input cell. The second type tabulates the effect on one formula of substituting pairs of values from two series into two specified input cells.

Table 11.2 Data Table 1 General Form

```
Blank cell      Formula A       Formula B      ..  Formula M
Value 1         Result   A1     Result   B1    ..  Result   M1
Value 2         Result   A2     Result   B2    ..  Result   M2
   :               :               :                   :
   :               :               :                   :
Value N         Result   AN     Result   BN    ..  Result   MN
```

Table 11.2 shows the general form of Data Table 1. In this case each value in turn is substituted into a input cell somewhere else on the spreadsheet and the result on the various formulae tabulated. For instance, Result MN is formed from the effect of Value N on Formula M. This technique is useful in sensitivity analyses.

For our crosstabulation we want to count the numbers of persons who fall into both of 2 categories, and so we use Data Table 2 as shown in table 11.3.

Table 11.3 Data Table 2 General Form

```
Formula         Value    A      Value    B     ..  Value    M
Value 1         Result   A1     Result   B1    ..  Result   M1
Value 2         Result   A2     Result   B2    ..  Result   M2
   :      :        :               :                   :
   :      :        :               :                   :
Value N         Result   AN     Result   BN    ..  Result   MN
```

In this case, when the Data Table command is issued each value in the left hand column is substituted into a first input cell (which you are asked to define), with all combinations of the values in the topmost row being substituted into a similarly defined second input cell. The results are automatically tabulated according to the effect of the substitutions on the formula in the top left of the table. For instance, Result MN is formed from the effect of Values M and N on the Formula.

In our example we shall use this type of Data Table with the @DCOUNT function (which we have already entered) as the formula, with the Criterion range, **CR**, and with two of its blank cells being defined as the input cells. In B35 enter the label **PART**, in C35 the label **FULL**, in A36 the label **M** and in A37 **F**. Then call up the menu and select Data Table 2 with a range of A35..C37. Define Input cell 1 to be D32, immediately below the **SEX:** label, and Input cell 2 to be E32, directly below the **TIME:** field name. This will cause **M** and **F** to be substituted in turn in the formula in combination with **PART** and **FULL**, so that the crosstabulation of sex by time is generated:

Table 11.4 Crosstabulation of Sex by Time

	A	B	C	D	E	F
35		26	PART	FULL		
36	M		5	9		
37	F		7	5		

Notice that, like other Data commands, Data Table has generated actual values rather than formulae. This means that if the original database is changed the table will no longer reflect it. However the table is easily recalculated by pressing the **[8/TABLE]** key.

11.4 Chi-squared Test

The Chi-squared test can be used to analyse any sample data in the form of frequency counts in particular categories, and it is commonly used with contingency tables of crosstabulated values as we have here. The original data comprise the **'observed'** values which can be compared with some **'expected'** ones.

The formula for the Chi-squared statistic is:

$$\chi^2 = \Sigma \{ (\text{Observed} - \text{Expected})^2 / \text{Expected} \} \qquad [11.1]$$
where
Expected = (Row Total * Column Total) / Table Total   [11.2]

In order to generate **expected** values it is necessary to form **row**, **column** and **table totals** for the raw data. These can easily be found using the @SUM function.

Table 11.5 Totals Calculated

	A	B	C	D	E	F
35		26	PART	FULL	Total:	
36	M		5	9	14	
37	F		7	5	12	
38	Total:		12	14	26	

125

Since we can break the Chi-squared test down into four stages, it makes sense to copy the observed table down four times to A40, A45, A50 and A55 respectively, and then to overwrite the data values. Notice that although you can only copy a range to one area when you issue the Copy command, when you have copied the table once you can then copy the original and the new table together to give a total of four tables.

The first new table, starting at A40, is the expected table. So the top left hand corner can be labelled **EXP**. The first expected value in cell B41 is generated by multiplying the appropriate **row** and **column totals** and then dividing by the **overall total** of the observed table. Since the **row totals** are in row 38, the number of this address needs to be absolute; whilst the column letter can be either B or C and is thus relative. Similarly the **column totals** are in column D, which should be absolute, with relative row reference. The **table total**, of course, is absolute in both row and column. Hence the resulting formula is:

+B$38*$D36/$D$38

This can then be copied over the rest of the **expected** table. The row, column and overall totals will change when the formula is entered in the first cell, but should revert to those in the **observed** table when copying is completed.

The next table is the **Observed** minus the **Expected**. Cell B46 has the formula:

+B36-B41

This can also be copied over the rest of this table. Now, however, the row, column and overall totals should become zero.

The next table to be formed is a squared version of the previous, so cell B51 contains:

+B46^2

Again this can be copied over the rest of the table. The final table is begun in cell B56 with the formula:

+B51/B41

This too can be copied over the rest of the final table. The overall total is the actual **Chi-squared statistic**.

The worksheet, formatted to 2 decimal places, should look like table 11.6.

So, in this example, the Chi-squared value is 1.33. The formula for the associated degrees of freedom is:

$$df = (\textbf{rows} - 1) * (\textbf{columns} - 1) \qquad [11.3]$$

which, in this case, is just 1. These values can be substituted into the worksheet presented in chapter 8 on statistical tables to find the associated significance probability. The higher the Chi-squared value obtained, the more likely that there is a relationship between the row and column criteria, that is between sex and time worked.

Table 11.6 Chi-squared Worksheet

	A	B	C	D	E	F
35	26	PART	FULL	Total:		
36	M	5	9	14		
37	F	7	5	12		
38	Total:	12	14	26		
39						
40	EXP	PART	FULL	Total:		
41	M	6.46	7.54	14		
42	F	5.54	6.46	12		
43	Total:	12	14	26		
44						
45	OBS-EXP	PART	FULL	Total:		
46	M	-1.46	1.46	0		
47	F	1.46	-1.46	0		
48	Total:	0	0	0		
49						
50	(O-E)^2	PART	FULL	Total:		
51	M	2.14	2.14	4.27		
52	F	2.14	2.14	4.27		
53	Total:	4.27	4.27	8.54		
54						
55	(O-E)^2/E	PART	FULL	Total:		
56	M	0.33	0.28	0.61		
57	F	0.39	0.33	0.72		
58	Total:	0.72	0.61	1.33		

11.5 Summary

The Chi-squared test is a non-parametric test, that is it does not need to assume that the input data values are drawn from populations that are normally distributed. This makes it a robust technique which can used in a wide variety of situations. It is not restricted to use with two-way tables, although they are commonly analysed with it. Using a spreadsheet enables the Chi-squared formula to be broken down into a series of simple steps, some of whose consistency can readily be demonstrated.

12
Regression and Correlation

- Range Name Labels
- Graph Name
- File Combine

12.1 Introduction

Regression and **Correlation** analysis is used when we have sets of measurements on two variables, **X** and **Y**, which we believe are related to one another. The first step in testing for the existence of such a relationship is to draw and inspect a scatter diagram of points, each of which represents a pair of observations of the variables **X** and **Y**. This is easily accomplished in 1-2-3, using an **XY** graph. If the scatter diagram suggests that there is indeed some relationship between the variables, we can estimate it and test its strength and reliability using regression and correlation analysis. The estimated relationship will enable us to predict the **Y** value for a given **X** value.

The methods we shall use in this chapter are applicable when the relationship between the variables takes the form of a straight line. The pattern of the points in the scatter diagram will then be approximately linear. If it is apparent that this pattern is non-linear, we may be able to apply a transformation to the measurements so that there is a linear relationship between the new variables and we can then apply the methods of this chapter.

Using our 1-2-3 spreadsheet we will be able to make our calculations using definitional formulae. We shall find Range Name useful in making these formulae more readable, and we shall create some of the names using the **Range Name Labels** command. Automatic recalculation will be useful in making predictions for various values, in demonstrating the least squares property of the regression line, and in investigating the influence of outliers.

12.2 Scatter Diagrams

- **Graph Name**

The data listed in table 12.1 give the quantity of rice produced in China and the number of tractors there for the period 1975-82. We shall investigate whether rice production **depends** on the number of tractors. For regression analysis it is important to distinguish the **dependent variable** or **effect**, which is always labelled **Y**, from the **independent variable** or **cause**, which is always labelled **X**. In our case it is rice production which we think is affected by the number of tractors, so we label rice production **Y** and the number of tractors **X**. When plotting a scatter diagram the dependent variable, **Y**, always goes on the vertical axis, and the independent variable, **X**, on the horizontal one.

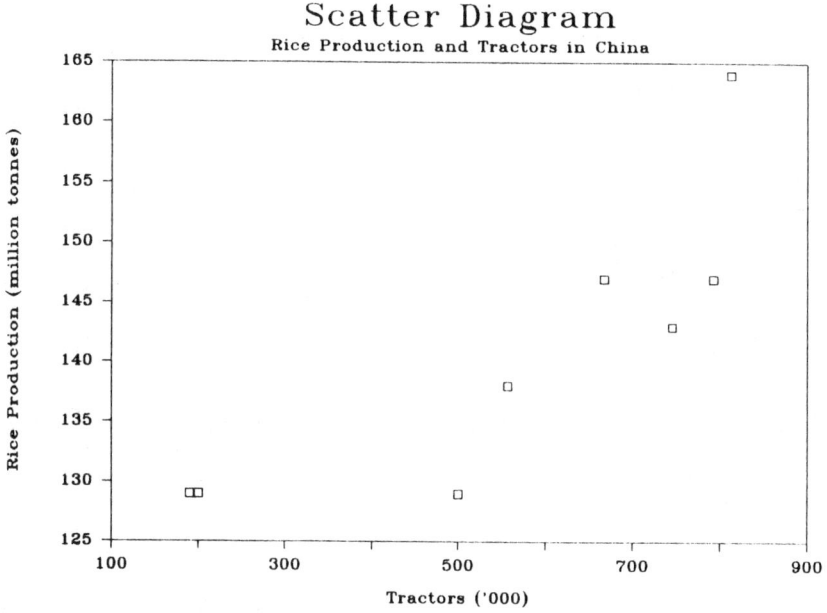

Fig. 12.1 Scatter Diagram

Enter the data from table 12.1 in the first three columns of your spreadsheet, then plot a scatter diagram using 1-2-3's Graph command. Select Type XY, define the **X** range as the tractor values and the A range as the corresponding amounts of rice produced. To title your graph and to get the data points plotted as symbols, as shown in fig. 12.1 you must use the Options command. Notice that using automatic scaling the origin is not shown on the graph. To title the graph itself and the axes use Options Titles, selecting each type of title in turn and typing the description you require. Options Format allows you to specify for data range **A** that you wish Symbols to be plotted.

Table 12.1 Data on Rice Production and Tractors

```
         A           B           C           D
1     Rice Production (million tonnes) and
2     Number of Tractors (thousands) in China
3     ----------------------------------------
4        Year      Rice     Tractors
5                   Y          X
6     ----------------------------------------
7        1975      129        190
8        1976      129        200
9        1977      129        500
10       1978      138        557
11       1979      147        667
12       1980      143        745
13       1981      147        792
14       1982      164        812
15    ----------------------------------------
16    Source: UN FAO, Production Yearbook
```

You will perhaps want to plot other graphs as you develop your present spreadsheet. 1-2-3 stores your latest graph as part of the spreadsheet information. To tell it to store your current graph so that it will still be available when you set up a new one you should use the Name command in the Graph sub-menu. If you wish to print your scatter plot you must also use the Save command in that sub-menu to save the graph for printing later when you have the 1-2-3 Print Graph program loaded in your machine. You may give the same name to your graph under both the Name and Save commands. 1-2-3 adds a suffix to your names to distinguish the different types of files.

The graph shows clearly that rice production has increased as the number of tractors has increased. It appears that the relationship between the two variables could be linear, and we shall proceed to estimate the regression line for these data points.

12.3 Regression Calculation

- ### Range Name Labels

The line whose equation we shall now calculate is called the **line of regression of Y on X**. Since we have defined our dependent variable, **Y**, to be rice production we shall be calculating the regression of rice production on the number of tractors in China, and the equation we obtain will provide an estimate of the relationship between the variables.

The line of regression takes the form

$$Y = a + bX \qquad [12.1]$$

where **a** the **intercept** on the vertical axis, and **b**, the **slope** or **gradient** of the line, are chosen using the least squares principle. That is, they are chosen so as to minimise the sum of the squares of the vertical distances of the points from the line. These vertical distances are called **residuals**. We shall make calculations involving them in section 12.6. The regression line is plotted with the data points in fig. 12.3.

Formulae for **a** and **b** are given in most statistics text books. Often that for **b** will be given in a form which makes its hand calculation easier, but we shall use the basic definitional form comprising sums of cross products and squares of deviations from the mean. The usual notation for the mean of **X** is $\bar{X}$, but it is not possible in 1-2-3 to overstrike a letter with a symbol, so we shall use the names **Xbar** and **Ybar** for the means of **X** and **Y** respectively, and we shall refer to them by these names in all our formulae.

The appropriate basic statistical formula, then, for **b**, the slope coefficient, is

$$b = \frac{\sum(X-Xbar)(Y-Ybar)}{\sum(X-Xbar)^2} \qquad [12.2]$$

where the summations are over whole sets of values. If we define

$$x = X - Xbar \qquad [12.3]$$

and

$$y = Y - Ybar \qquad [12.4]$$

we may write

$$b = \frac{\sum xy}{\sum x^2} \qquad [12.5]$$

Another formulation is to divide both numerator and denominator of 12.2 by n, the number of pairs of observations.

This gives

$$b = \frac{\{\sum(\textbf{X-Xbar})(\textbf{Y-Ybar})\}/n}{\sum(\textbf{X-Xbar})^2/n} = \frac{\text{Covariance}(\textbf{X,Y})}{\text{Variance }(\textbf{X})} \qquad [12.6]$$

By definition, the numerator of **b** in 12.6 is called the **covariance** of **X** and **Y** and the denominator is the **variance** of **X**. The variance of **X** is the square of the standard deviation of **X** when the latter is defined, as it is in 1-2-3, using **n** as divisor.

Table 12.2 Deviations from Means

```
            A           B           C           D           E
1    Rice Production (million tonnes) and
2    Number of Tractors (thousands) in China
3    ----------------------------------------------------
4           Year       Rice    Tractors    Y-Ybar      X-Xbar
5                       Y          X          y           x
6    ----------------------------------------------------
7           1975       129        190      -11.75     -367.88
8           1976       129        200      -11.75     -357.88
9           1977       129        500      -11.75      -57.88
10          1978       138        557       -2.75       -0.88
11          1979       147        667        6.25      109.13
12          1980       143        745        2.25      187.13
13          1981       147        792        6.25      234.13
14          1982       164        812       23.25      254.13
15   ----------------------------------------------------
16   @SUM             1126       4463        0.00        0.00
17   @AVG           140.75    557.875
18                   Ybar        Xbar
19   @STD            11.43     232.67
20                    sdY         sdX
21   @COUNT             8
22                      n
```

The least squares principle also yields a formula for the intercept term, **a**, in our regression equation.

It is defined as

$$\textbf{a = Ybar - b.Xbar} \qquad [12.7]$$

This ensures that the point (**Xbar, Ybar**), the point representing the mean values of the data sets, lies on the regression line.

Tables 12.2 and 12.3 show the spreadsheet calculations for **b** using both formulae 12.5 and 12.6, and for **a** using 12.7. The @AVG and @STD functions are used to find the means and standard deviations of the **Y** and **X** values. Labels indicating the names these are to be given **Ybar, Xbar, sdY** and **sdX**, are entered in the cells immediately below each of them. Similarly the @count function is used to find the number of observations, **n**, and a label is entered below it. The command **Range Name Labels Up** can then be used to assign the names displayed in the labels to the cells immediately above them. The appropriate range for this command will include all the labels which are to be assigned as names.

Table 12.3 Squared Deviations from Means

	D	E	F	G	H
1					
2					
3	------	------	------	------	------
4	Y-Ybar	X-Xbar			
5	y	x	y*y	x*x	x*y
6	------	------	------	------	------
7	-11.75	-367.88	138.06	135332.02	4322.53
8	-11.75	-357.88	138.06	128074.52	4205.03
9	-11.75	-57.88	138.06	3349.52	680.03
10	-2.75	-0.88	7.56	0.77	2.41
11	6.25	109.13	39.06	11908.27	682.03
12	2.25	187.13	5.06	35015.77	421.03
13	6.25	234.13	39.06	54814.52	1463.28
14	23.25	254.13	540.56	64579.52	5908.41
15	------	------	------	------	------
16	0.00	0.00	1045.50	433074.88	17684.75
17			ySS	xSS	Sumxy
18					
19				Sumxy/n	2210.593
20					Cov
21	Sumxy/@SQRT(ySS*xSS)		0.831	Cov/(sdY*sdX)	0.831
22					r
23				r*r	0.691
24					rsq
25		Sumxy/xSS	0.041	Cov/(sdX*sdX)	0.041
26					b
27				Ybar-b*Xbar	117.969
28					a

Columns can then be constructed containing deviations of each set of values from the appropriate mean, namely **Y-Ybar**, or **y**, from equation 12.4 and **X-Xbar**, or **x**, from 12.3. Columns of squares of each of these, **y*y** and **x*x**, and their cross products, **x*y**, can be formed also. Notice that the addresses used in the formulae entered at the top of each column should be relative, except for those of the means, which must be made absolute. You can refer to the means by name, indicating that absolute addresses are required by preceding the names with a [$] sign. That is, you should type **$Ybar** and **$Xbar** as appropriate. You can enter the formulae for all the five columns you wish to construct, then Copy from the range containing these five formulae down to the bottom row of the table. The other values in the table are thus all computed and entered simultaneously. Notice that to improve the readability of tables 12.2 and 12.3, fixed formats have been chosen and some column widths have been adjusted.

The transformation from variables **X** and **Y** to variables **x** and **y**, as specified in expressions 12.3 and 12.4 and performed in the construction of the x and y columns, is called a **change of origin**. The origin for each variable is now its mean, so that where the original variables were greater than the mean the new variables are positive but where the original variables were below the mean the new variables are negative. A scatter diagram for the transformed data is shown in fig. 12.2. Notice that the scatter of points has the same pattern as in fig. 12.1, but that the axes now have their origin at what was the point (**Xbar, Ybar**) on the original scatter plot. Fig 12.2 was produced by graphing the y values as the A range against the **x** values as the X range of an XY graph.

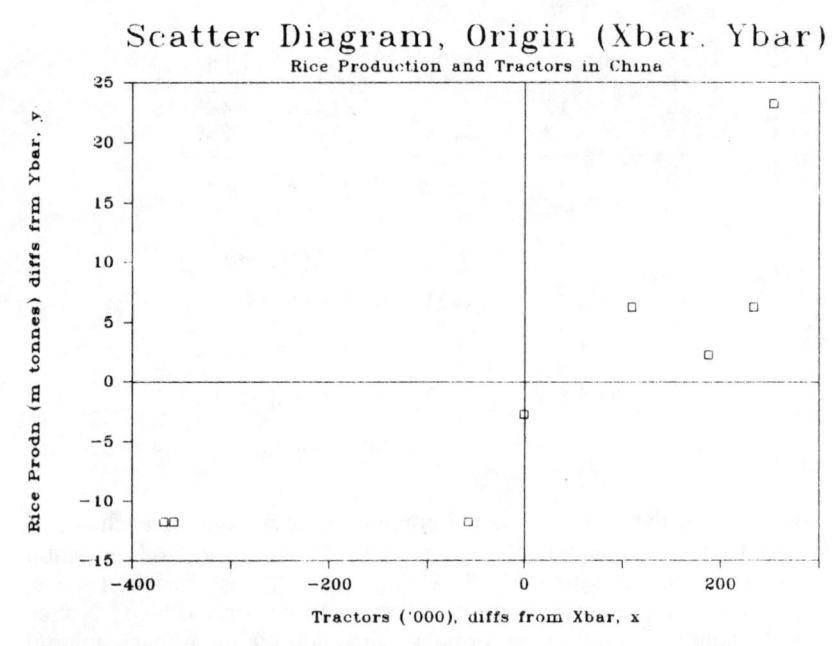

Fig. 12.2 Scatter Diagram, Origin (Xbar, Ybar)

The @SUM function can then be used, entering it once and copying it to total each of the columns. Names may then be given to those totals which are to be used in computations, by entering labels containing the names and assigning them as before. Because there is no summation sign in 1-2-3 and we cannot superscript a 2 to indicate a squared value, the following conventions have been adopted in the names assigned: **SS** at the end of a name indicates a sum of squares, and **Sum** at the beginning indicates a summation. Remember that although we may type capital or lower case letters in names, 1-2-3 does not distinguish between them.

If Σxy is named **Sumxy** and Σx^2 **xSS**, you can compute **b** using formula 12.5 by entering

    +Sumxy/xSS

A label indicating that this formula has been used is entered in table 12.3 in the cell to the left of that in which the calculation takes place. Notice that the computational formula in 1-2-3 should begin with a [+] to indicate a value. Labels displaying the formulae used improve the readability of a spreadsheet; more will be entered as we perform further calculations. Readability is further improved by displaying calculated values in a fixed format. For the expressions we are now evaluating, three decimal places are perhaps appropriate.

Formula 12.6 computes **b** as the quotient of the covariance divided by the variance of **X**. We can calculate the covariance as **Sumxy/n**; remember to precede this with a [+] in 1-2-3. Table 12.3 shows the covariance named **Cov**. The denominator of expression 12.6, the variance of **X**, is formed as the square of the standard deviation of **X**, **sdX**, which we have already calculated. The regression slope, b, can then be computed by entering

    +Cov/(sdX*sdX)

Notice that the parentheses are required when the instruction is entered in this form. Division and multiplication are operations which have the same precedence, and are therefore carried out from left to right. Without the parentheses, then, the quotient of **Cov/sdX** would be multiplied by **sdX**; the parentheses ensure it is further divided by **sdX** as required. Using formula 12.6, therefore, you can calculate **b** using the @STD function instead of forming a column of **x*x** values.

When you have named one of your **b** values appropriately you can calculate a according to formula 12.7 by entering

    +Ybar-b*Xbar

Using the results displayed in table 12.3 you can now write down your line of regression:

$$Y = 117.969 + 0.041 \, X \qquad [12.8]$$

This equation indicates that if there were no tractors in China (**X** = 0), rice production would be 117.969 million tonnes. It is possible, however, that your equation may not be valid for **X** values outside the range of those for which it has been fitted. The slope coefficient gives a description of what might happen if the number of tractors were increased (or decreased). For every 1000 extra tractors (**X** increases by 1) we expect rice production to increase by 0.041 million, or 41,000 tonnes.

The fitted regression line is shown together with the scatter of points in fig. 12.3. We shall see how to add the regression line to our scatter diagram in section 12.5, below.

12.4 Correlation Coefficient

The correlation coefficient, **r**, measures the strength of the linear association between two variables. If there is no relationship at all its value is zero, while if the points all lie exactly on a line its value is 1 when the line slopes upwards, or -1 when it slopes downwards. For n pairs of values of **X** and **Y** the correlation coefficient is defined as

$$r = \frac{\sum(X - \bar{X})(Y - \bar{Y})}{\sqrt{[\sum(X-\bar{X})^2 \sum(Y-\bar{Y})^2]}} \qquad [12.9]$$

Using our earlier notation this can be rewritten as

$$r = \frac{\sum xy}{\sqrt{[\sum x^2 \sum y^2]}} \qquad [12.10]$$

or, dividing numerator and denominator by **n**, so that the expression within the square root is divided by n^2, we have

$$r = \frac{\sum xy/n}{\sqrt{[(\sum x^2/n)(\sum y^2/n)]}} = \frac{\text{Covariance }(X,Y)}{\text{Std}(X)\,\text{Std}(Y)} \qquad [12.11]$$

Both expressions 12.10 and 12.11 have been used in table 12.3 to calculate **r**. For the former you should ensure $\sum y^2$ is named **ySS** and enter

+Sumxy/@SQRT(xSS*ySS)

Computation by the second method is easier, since the sums of the squared deviations are not required. You simply enter

```
+Cov/(sdX*sdY)
```

The value twice displayed in table 12.3 for the correlation coefficient is 0.831. One of these values has been labelled and named **r**. The square of this value, **rsq**, has been calculated also using the 1-2-3 formula

```
+r*r
```

It is shown in table 12.3 as 0.691. Notice that we are using **sq** at the end of the name to indicate a squared variable. The value of **rsq** tells us that 69.1% of the variation in rice production, **Y**, in our data is accounted for by the regression relationship we have fitted. Remember, however, that this statistical relationship does not prove the existence of a **causal** relationship between the variables.

12.5 Prediction

- **Graph Name Use**

The regression line we have fitted

$$Y = 117.969 + 0.041X \qquad [12.8]$$

is an estimate of the relationship between **X** and **Y**. We can use it to predict **Y** values corresponding to various **X** values of interest substituted into the regression equation in turn. This is done for our original **X** values in the first column of table 12.4. The values in that column, which is labelled **Ypred**, are computed by entering and copying the formula

```
$a+$b*C7
```

Notice that [$] signs are required to make the references to the cells containing the regression coefficients absolute, and notice also that 1-2-3 interprets a [$] sign as beginning a value. The first **X** value, shown in table 12.2 as 190, is held in cell C7. Multiplication has a higher priority in 1-2-3 than addition, so the **X** values will each be multiplied by the value of **b** before the value of **a** is added.

Table 12.4 Predicted Values and Residuals

```
                I          J              K          L         M
 1
 2         Predicted    Residual
 3         ---------------------------------------------------------
 4           a+b*X      Y-Ypred
 5           Ypred         e             e*e        X*X
 6         ---------------------------------------------------------
 7          125.73        3.27          10.71      36100
 8          126.14        2.86           8.20      40000
 9          138.39       -9.39          88.11     250000
10          140.71       -2.71           7.37     310249
11          145.21        1.79           3.22     444889
12          148.39       -5.39          29.07     555025
13          150.31       -3.31          10.96     627264
14          151.13       12.87         165.71     659344
15         ---------------------------------------------------------
16         1126.00         .00         323.34    2922871
17                                       eSS       XaSS
18
19                                    eSS/(n-2)   53.890
20                                                sigsq
21                    @SQRT(sigsq)                 7.341
22                                                sigma
23            800                     a+b*Xo     150.637
24            Xo                                 YpredXo
25                                    Xo-Xbar    242.125
26                                                Xdiff
27        @SQRT(sigsq*(1+1/n+Xdiff^2/xSS))         8.241
28                                                SEpred
29                    @SQRT(sigsq/xSS)             0.011
30                                                 SEb
31        @SQRT(XaSS*sigsq/(n*xSS))                6.743
32                                                 SEa
33            n-2           6          b/SEb       3.661     5.559
34                                                  t_b      t_a
```

The regression line we have fitted can now be drawn on our earlier scatter diagram. The last graph you defined is stored by 1-2-3. A graph which has previously been named (note, not saved - that is for printing) can be recalled by the **Graph Name Use** command. With your scatterplot in the original units as the current graph, define the **B** range to be the predicted values you have just calculated. Use Options to Format that range as a line, and you will obtain the graph shown in fig. 12.3.

If you print this graph you will be able to read from it, approximately, the **Y** values on the regression line corresponding to the **X** values. These are the predicted Y values, **Ypred**, that we have calculated.

We may wish to predict a **Y** value corresponding to some other **X** value, say **Xo**. A value of 800 for **Xo** has been entered in the spreadsheet shown in table 12.4 below the column of values we have just calculated. To the right of **Xo** the formula

+a+b*Xo

has been entered to calculate the corresponding predicted **Y** value, **YpredXo**. On re-entering a new value for **Xo** in the same cell, the new predicted **Y** value will automatically be recalculated.

Scatter Diagram with Regression Line
Rice Production and Tractors in China

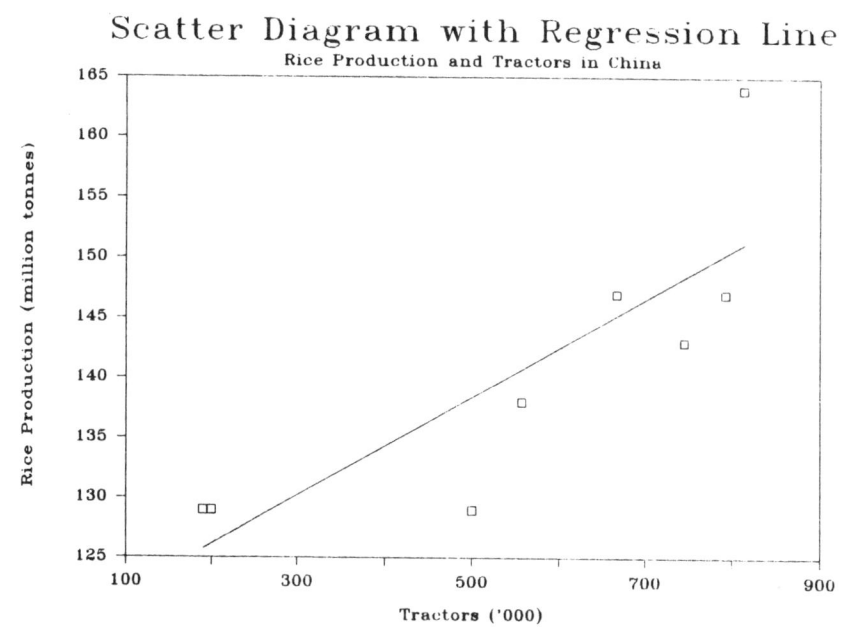

Fig. 12.3 Scatter Diagram with Regression Line

Your main interest in fitting a regression line may be to use it for forecasting. In this example you may wish to forecast the effect on rice production of further investment in tractors. Clearly, we can make **Xo** as large as we like and the equation will predict a corresponding **Y** value for rice production. It may not be meaningful, however, to do this. Other conditions may also change, so that the regression line may no longer be valid.

2.6 Residuals and Standard Errors

- **File Combine Copy**

We have described our regression equation as **estimating** the linear relationship between **X** and **Y**. This implies that the equation is subject to possible error, both in the estimate of **b** and in that of **a**. The calculation of standard errors enables us to make probability statements about the values of these coefficients and about predictions from the regression line.

We shall begin by calculating the **standard error of estimate**, which is often denoted σ, but which we shall name **sigma**.

It measures the spread of the **residuals**, labelled **e**, which are the differences between the actual and predicted **Y** values, **Y - Ypred**. That is,

$$\mathbf{e = Y - Ypred} \qquad [12.12]$$

In diagrammatic terms, the residuals are the vertical distances of the points shown in fig. 12.3 from the fitted regression line. Points above the line have positive residuals, points below it negative ones. The standard error of estimate, **sigma**, is then defined as

$$\mathbf{sigma} = \sqrt{\left[\sum (\mathbf{Y - Ypred})^2 / (n-2)\right]} \qquad [12.13]$$

$$= \sqrt{\left[\sum e^2 / (n-2)\right]} \qquad [12.14]$$

To calculate **sigma** we form columns containing the values of **e** and of the squares of these values, **e*e**. When forming **e**, each predicted **Y** value is to be subtracted from the corresponding actual **Y** value, so relative addresses are appropriate, as they are also when forming **e*e**. Having constructed these columns we sum them and notice that the sum of the residuals, $\sum \mathbf{e}$, is zero. This is a property of the least squares fit of the regression line. In diagrammatic terms it implies that the vertical distances from the line of points lying above it are balanced by those of points lying below.

The sum of the squared residuals, $\sum e^2$, is named **eSS**. It is this sum of squares which is minimised by the formulae chosen for the regression coefficients **a** and **b**.

The square of the standard error of estimate, which is usually denoted σ^2, but which we shall name **sigsq**, can be calculated by entering the formula

> +eSS/(n-2)

Using 12.14, then, **sigma** is obtained with

> @SQRT(**sigsq**)

The standard errors of **b** and of **a**, which we shall name **SEb** and **SEa** respectively, are defined as follows:

> **SEb** = $\sqrt{[\text{sigsq}/\sum x^2]}$ [12.15]

and

> **SEa** = $\sqrt{[\text{sigsq} \cdot \sum X^2 / (n \sum x^2)]}$ [12.16]

The first of these we can calculate immediately using the formula

> @SQRT(**sigsq/xSS**)

Its value is 0.011.

To calculate **SEa** we must first form a column containing the squares of the original **X** values. This is shown in table 12.4 headed **X*X**. We form the sum of this column and give it a name, but since **XSS** would be indistinguishable in 1-2-3 from **xSS** which has already been used as a name, the new sum of squares has been named **XaSS** in table 12.4. We can now form the standard error of **a**, **SEa**, by entering

> @SQRT(**XaSS*sigsq/(n*xSS)**)

This gives a value of 6.743.

The standard errors can be used to find confidence limits for **a** and **b** and to test hypotheses about their values. The appropriate distribution to use for making statements about probabilities of the values of **a** and **b** is the **t** distribution with **n**-2 degrees of freedom.

Table 12.4 shows the calculation of **t** values to test whether **b** and **a** (separately) are significantly different from zero, together with the number of degrees of freedom for the **t** distribution. For **b** the test statistic, labelled **t_b**, is **b/SEb**, and it has the value 3.661.

To conduct the test you can refer to **t** tables with (**n**-2) = 6 degrees of freedom, which give $t_{0.99}$ = 3.14. Since **t_b** is greater than this, we can reject at the 0.01 significance level the hypothesis that **b** is in fact 0. Another approach is to calculate the significance value of your test statistic, using the spreadsheet for continuous probability calculations which we set up in chapter 8. To do this, position the cell pointer in a blank area of your spreadsheet and use the File Combine Copy Entire File command to bring the spreadsheet shown in table 8.7 into your current spreadsheet. Ensure the cell containing the Numerator DF is blank, enter the code for the **t** distribution (2), the degrees of freedom (6), and the **t** value (3.661). The probability will be calculated as 0.005, which is, of course, less than the required significance probability of 0.01. We conclude, then, that tractors and rice production really are positively related to one another. Applying a similar test to **t_a** also leads us to decide that **a** is really not zero. This implies that even if there were no tractors, there would be some rice production.

We have used our regression line to predict a value, **YpredXo**, corresponding to a particular value **Xo**. But since our regression line is subject to estimation error, and since any individual observation might not lie exactly on the regression line, our prediction is also subject to error. The size of this error depends on how far **Xo** is from **Xbar**, so we begin by calculating **Xo-Xbar** and naming the result **Xdiff**. The **standard error of prediction, SEpred** is defined as

SEpred = √{**sigsq**[1 + 1/n + (**Xo-Xbar**)²/∑x²]} [12.17]

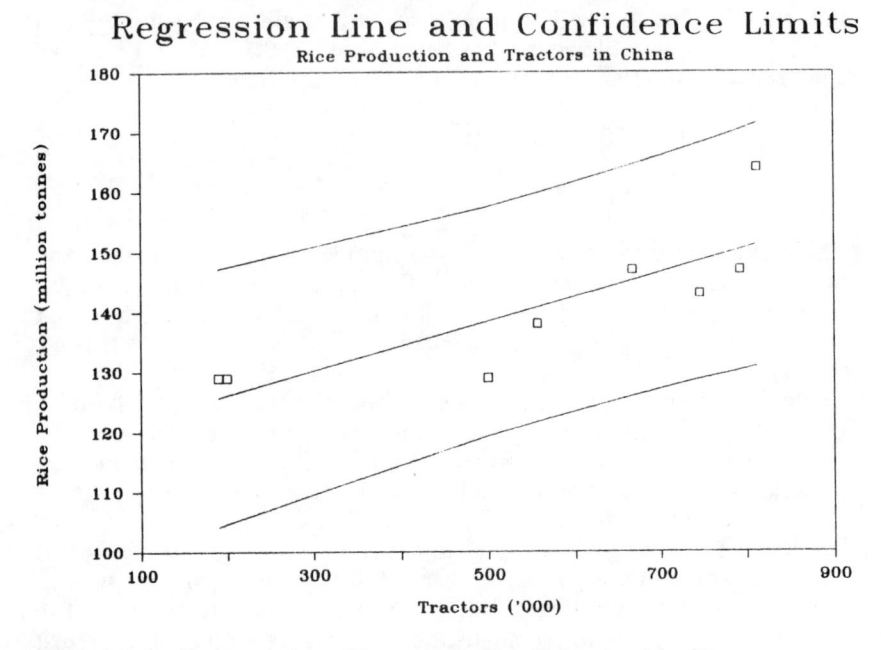

Fig. 12.4 Regression Line and Confidence Limits

The formula required to calculate this in 1-2-3 is

@SQRT(sigsq*(1+1/n+Xdiff^2/xSS))

Table 12.4 shows that for **Xo** = 800 the standard error of prediction is 8.241. This standard error can be multiplied by the appropriate t value to set confidence limits for the prediction. If we do this for various values of **Xo** we obtain curvilinear limits, which become wider the further away **Xo** is from **Xbar**.

You might like to construct confidence limits for each of your original **X** values. Form a column of values of **SEpred** multiplied by 2.4469 (the **t** value for 6 degrees of freedom at the 95% probability level), then columns containing these values added to and subtracted from the **Ypred** values, to form the lower and upper confidence limits. Define the last two columns as data ranges C and D of the graph shown in fig. 12.3 and view your graph again. You should obtain the graph pictured in fig 12.4, which shows slightly curved 95% confidence limits above and below the regression line.

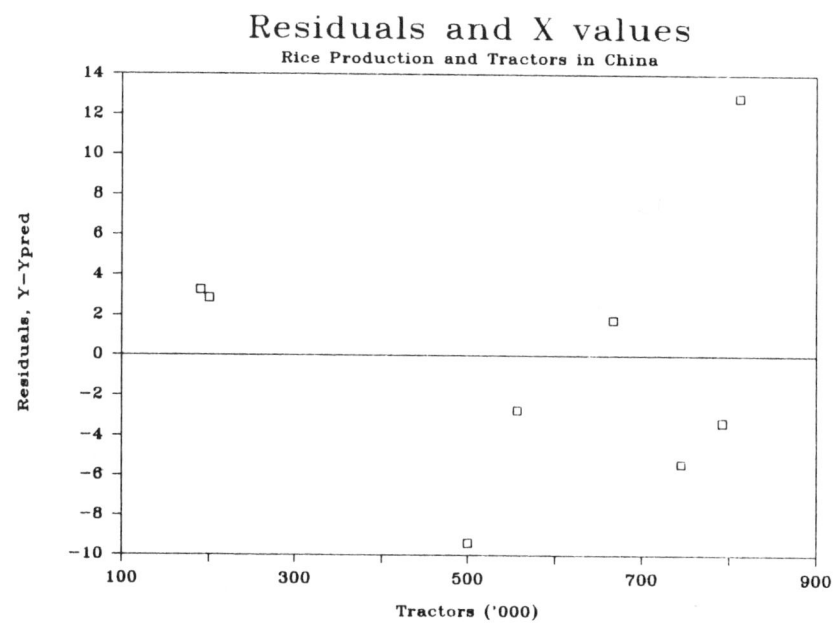

Fig. 12.5 Residuals and X Values

The least squares estimators, **a** and **b**, which we have calculated, can be shown to have certain optimal statistical properties provided that the residuals conform to certain assumptions which are made. With time series data it is useful to plot the residuals, **e**, to examine whether they appear to be random, and also whether there seems to be any relationship between the residuals and the value of **X**. A graph of the residuals against the X values is shown in fig. 12.5.

When you have completed your spreadsheet be sure to save it before making any of the changes suggested in the next two sections.

12.7 Least Squares Property

We stated in the previous section that the formulae chosen for **a** and **b** are those that minimise the sum of squared residuals Σe^2, or eSS in our spreadsheet.

To demonstrate that this is so, note the value of **eSS**, then try entering various values in turn into the cell labelled **b**. All values calculated from **b** will recalculate, so **a**, the predicted **Y**'s and the residuals will take on different values. Note the various new values of **eSS**. You will find they are all larger than 323.34, unless you enter a value of 0.041 for **b**. For example, if you enter 0.06 for **b**, **a** is then 107.278 and **eSS** is 482.40. With 1-2-3 you could in fact use a data table to investigate which pair of values for **b** and **a** give the smallest value of **eSS**.

12.8 Influence of Outliers

The recalculation powers of 1-2-3 make it easy to investigate the sensitivity of your regression equation to values which might be regarded as outliers. Amongst our data, the observations for 1975 and 1976 are thought to be possibly unreliable. You might like to substitute different values for **Y** and **X** in these years. We, however, shall eliminate these observations from our analysis.

The values for 1976 may be removed using Worksheet Delete Row, and the ranges over which the columns are summed will be adjusted accordingly. We cannot, however, eliminate the 1975 values in this way, because they form end points of the ranges summed and 1-2-3 will display an error message. We may remove them instead using Range Erase. The resulting row of blank entries will be included in the various sums without any detriment. Table 12.5 shows the revised version of parts of the spreadsheet shown in tables 12.2 and 12.3.

Table 12.5   1975 and 1976 Values Removed

```
         A          B           C              F          G          H
 1   Rice Production (million tonnes) and
 2   Number of Tractors (thousands) in China
 3   ------------------------------------------------------------------
 4   Year        Rice      Tractors
 5                Y           X               y*y        x*x        x*y
 6   ------------------------------------------------------------------
 7   1975
 8   1977        129         500            245.44   31981.36    2801.72
 9   1978        138         557             44.44   14843.36     812.22
10   1979        147         667              5.44     140.03     -27.61
11   1980        143         745              2.78    4378.03    -110.28
12   1981        147         792              5.44   12806.69     264.06
13   1982        164         812            373.78   17733.36    2574.56
14   ------------------------------------------------------------------
15   @SUM        868        4073            677.33   81882.83    6314.67
16   @AVG     144.67      678.83              ySS        xSS       Sumxy
17              Ybar        Xbar
18   @STD      10.62      116.82                     Sumxy/n    1052.444
19               sdY         sdX                                     Cov
20   @COUNT       6                        Cov/(sdY*sdX)          0.848
21                n                                                    r
22                                                      r*r        0.719
23                                                                  rsq
24                                        Cov/(sdX*sdX)           0.077
25                                                                    b
26                                          Ybar-b*Xbar          92.316
27                                                                    a
```

Notice that the line is now steeper (**b** has increased to 0.077) but that the intercept is less (**a** has decreased to 92.316). The residuals and standard errors will have changed also, and 2 degrees of freedom have been lost, so a different **t** distribution will apply.

12.9 Summary

A spreadsheet with graphical facilities is a very useful medium for simple regression and correlation analysis. It allows you to plot your data, your regression line, the confidence limits and the residuals. This lets you inspect visually the fit of your regression model.r

The calculations required to produce the regression coefficients, the residuals and the standard errors involve the repetition of simple arithmetic operations. Their hand calculation would be very tedious, but in a spreadsheet they can be carried out easily, one stage at a time.

To make your spreadsheet readable it is useful to name various values used in further calculations, and to display as labels the formulae used, which incorporate these names.

13
Analysis of Variance

13.1 Introduction

Analysis of variance, ANOVA, is a widely used and versatile hypothesis testing technique. It can be used to test whether one, or more than one factor, or a combination of factors, influences the quantity of output produced. The output being measured might be, for example, the amount of a product manufactured, or the yield of a crop. Factors whose influence you might wish to test are, for the first: the machine on which the product is produced, the operative employed, and the raw materials mix; and for the second: the level of fertiliser applied, the type of ground and the variety of seed.

We shall see in this chapter how to set up spreadsheet calculations to test whether one factor influences output, and also to test whether either of two factors separately influences output. More complex models can be developed, as has been indicated above, but they are beyond the scope of this book.

13.2 One Way Analysis of Variance

The **one way** analysis of variance model is applicable for testing whether just **one** factor has a real influence on the output variable being measured. If other factors may also influence the output variable but they are not included in the model, we must randomise their possible effects by the way we select our observations.

To test whether output differs with the production process used, the one way model attempts to separate the variation in the output observations into two parts. These are: that which is attributable to differences **between** production processes and, secondly, that which is due to chance and which occurs **within** all production processes. From the sums of squares which measure the variation of each type, two estimates of the population variance are formed, and these are compared. If the "between production processes" variance estimate is large by comparison with that attributed to chance and measured "within processes", then we shall conclude that the production process used really does affect output. That is, we shall decide that the outputs produced by different processes differ from one another by more than they would if the variation were due only to chance.

More formally, if the ratio, **F**, of our "between" to "within" variance estimates is sufficiently large, we shall reject the null hypothesis, H_0, which states that the mean output, μ, is the same for all production processes. We are testing, then,

$$H_0 : \mu_1 = \mu_2 = \mu_3 = ... = \mu$$

against the alternative hypothesis, H_1, that the means are not all equal. The test requires that the observations should be drawn from Normally distributed populations which have the same variance. We calculate Sums of Squares, **SS**, degrees of freedom, **df**, Mean Sums of Squares, **MSS**, and the **F** ratio; as shown in the Analysis of Variance table 13.1.

Table 13.1 Formulae for One Way Analysis of Variance

Source	SS	df	MSS	F
Between	$c\sum(Xbar-G/N)^2$	$r-1$	Between SS/$(r-1)$	Between MSS
Within	$\sum\sum(X-Xbar)^2$	$r(c-1)$	Within SS/$(c-1)r$	Within MSS
Total	$\sum\sum(X-G/N)^2$	$rc-1$		

where:

Xbar = $\sum X/c$ are the means of **c** observations for each process

G = $\sum\sum X$ is the grand total of observations

N = **rc** is the total number of observations

G/N = $\sum\sum X/rc$ = $\sum Xbar/r$ is the grand mean of observations.

Let us use this method to test whether there is any difference between the strengths of products produced using three different production processes. Table 13.2 shows the breaking strengths, in tonnes, of concrete beams manufactured using 3 different processes. Five randomly chosen observations are available for each process.

Table 13.2 Observations, X, for 3 Processes

```
          A   B   C     D     E     F     G   H  I     J      K
 1        |                                    |       (Xbar-G/N)^2
 2   Process|       Strength, X                |Total Xbar BETWEEN
 3        |-------------------------------------|-------------------
 4        1 |   8    10     8     8     7 |   41  8.2    3.74
 5        2 |  12    14    10    13    12 |   61 12.2    4.27
 6        3 |   9     9    10    11    11 |   50   10    0.02
 7        |-------------------------------------|-------------------
 8        |Rows  Cols         Total Obs |  152 Total  8.026
 9   Count|  3     5              15    |       G
10        |  r     c              N     |10.13  Avg
11        |                              |       G/N          40.13
12        |TOTAL, (X-G/N)^2             |           BETWEEN SS
13        |-------------------------------------|[N.B. Column sum
14        |  4.55  0.02  4.55  4.55  9.82| is multiplied by
15        |  3.48 14.95  0.02  8.22  3.48| count of columns]
16        |  1.28  1.28  0.02  0.75  0.75|
17        |                              |
18        |  9.32 16.25 4.586 13.52 14.05|57.73
19        |                              |TOTAL SS
20        |WITHIN, (X-Xbar)^2           |
21        |-------------------------------------|
22        |  0.04  3.24  0.04  0.04  1.44|
23        |  0.04  3.24  4.84  0.64  0.04|
24        |    1     1     0     1     1|
25        |                              |
26        |  1.08  7.48  4.88  1.68  2.48| 17.6
27        |                              |WITHIN SS
28        |ANALYSIS OF VARIANCE TABLE   |
29        |                              |
30        |    SS     df    MSS F RATIO |
31        |-------------------------------------|
32   Between|40.13    2  20.06  13.68   |
33   Within |17.60   12   1.47           |
34        |                              |
35   Total |57.73   14                   |
```

Enter the data into your spreadsheet as shown, together with appropriate labels. Counts are made of the numbers of rows, **r**, columns, **c** and observations, **N**. These are displayed below the data in table 13.2. We then form the columns headed **Total** and **Xbar** in table 13.2 using @SUM and @AVG respectively. These contain the totals and means of the values in each of the rows. The row totals are designated ΣX in the statistical formulae of table 13.1. The total of the totals gives the grand total, **G**, and the average of the averages gives the grand mean, **G/N**.

Appropriate identifying labels are entered below the various values that will be used in calculations. Except for **G/N**, which would be confused in formulae with an instruction to divide **G** by **N**, these labels should be assigned as names to the values above them, using Range Name Labels Up.

The between processes sum of squares is defined as

$$\text{BETWEEN SS} = c\sum(\textbf{Xbar} - \textbf{G}/\textbf{N})^2$$

Let us form a column containing the (**Xbar-G/N**)^2 values. The **Xbar** values must be identified by their individual addresses, but **G** and **N** have been assigned as names and so can be used in the formula. So that the formula can be copied down the column, both names must be made absolute. Summing this column and multiplying the result by the number of data columns, **c**, gives the BETWEEN SS shown, which is then named as such. Notice that range names may contain spaces.

In the bottom section of the spreadsheet shown in table 13.2, we now form a sub-table showing values of the squared deviations of the observations, **X**, from the overall mean, **G/N**. This allows us to calculate the total sum of squares, defined as

$$\text{TOTAL SS} = \sum\sum(\textbf{X} - \textbf{G}/\textbf{N})^2$$

A formula is entered in one corner of the sub-table to calculate a value of (**X** - **G/N**)^2. Judicious use of relative and absolute addressing enables us to copy this formula to form the other values of the sub-table. Summing the columns and then the column totals gives the TOTAL SS shown, which is also named.

We could now easily complete the Analysis of Variance table. One of the advantages of spreadsheets, however, is that they offer the possibility of consistency checking, and we shall avail ourselves of this opportunity by forming another sub-table to calculate the within processes sum of squares.

This sum of squares is defined as:

$$\text{WITHIN SS} = \sum\sum(\textbf{X} - \textbf{Xbar})^2$$

Again an appropriate formula for (**X** - **Xbar**)^2 can be copied to form the whole table of values. Notice that cell addresses must be used, and that the cell reference for Xbar must be absolute in the column but relative in the row for copying to be possible. We can now sum the columns and then the column totals to form the WITHIN SS, name it and check that the theoretical result holds that

$$\text{BETWEEN SS} + \text{WITHIN SS} = \text{TOTAL SS}$$

We now bring these sums of squares into an Analysis of Variance table, using formulae consisting of [+] followed by the appropriate name to enter the results from the cells where they are already shown. The degrees of freedom are entered utilising the values of **r** and **c**; and the mean sums of squares, **MSS**, are calculated by dividing both the between and within sums of squares by the appropriate numbers of degrees of freedom. These mean sums of squares are estimates of the population variance, and their ratio, calculated in the next column, is an **F** value which will be close to 1 if the null hypothesis is true. If, however, the means for the processes are not all equal, the **F** ratio will be greater than 1 and we can find its significance level by File Combining the spreadsheet shown in table 8.7 and making appropriate entries. If it is significant at the level required, we reject the null hypothesis.

Our calculated value of 13.68 is significant at the 0.0011 significance level. This very low value is the probability that, if there were no differences between the processes, we would have obtained sample values yielding an **F** value as high as the one we have calculated. We therefore reject the null hypothesis and conclude that the mean strengths of the beams are not all equal for the different production processes. Further statistical analysis is needed to pinpoint where the difference or differences lie.

The spreadsheet should recalculate automatically if data values are changed. You might like to try some "What if?" calculations.

13.3 Two Way Analysis of Variance

In **two way** analysis of variance we separate from the total variation the amounts which are attributable to each of **two** factors. The remaining variation is the within factors or chance variation. From these three parts of the total variation mean sums of squares are calculated, giving three variance estimates. Each of the two factor estimates is compared with the within factors estimate in an **F** test. If the **F** ratio for either factor is sufficiently large, we conclude that that factor really influences output.

Table 13.3 shows a spreadsheet while table 13.4 gives the basic statistical formulae for our computations.

Table 13.3 Two Way Analysis of Varience

```
     A      B     C       D      E      F     G     H       I       J
1    |            Output, X                  |           (Xibar-G/N)^2
2    Machine|Man 1  Man 2  Man 3  Man 4      |Total  Xibar   BETWEEN
3       1 |   2.0    2.5    3.0    2.3       |  9.8  2.450   0.412
4       2 |   3.2    2.7    3.5    3.0       | 12.4  3.100    .000
5       3 |   3.5    3.6    4.0    3.8       | 14.9  3.725   0.401
6    -------|-------------------------------- |-------------------
7    Total|   8.7    8.8   10.5    9.1       | 37.1  TOTMC   0.813
8         |                                   |       G
9    Xjbar| 2.900  2.933  3.500  3.033       | 3.092  Avg   TOTMC*c
10        |                                   |       G/N           3.252
11        |      Rows   Cols          Obs    |       BET MACHINES SS
12        |        3      4            12    |
13        |        r      c             N    |
14   BET MEN SS  (Xjbar-G/N)^2               |TOTMN   TOTMN*r
15        |  0.037  0.025  0.167  0.003     | 0.232   0.696
16        |                                   |       BET MEN SS
17        |TOTAL, (X-G/N)^2                  |
18   -------|--------------------------------
19        |  1.192  0.350  0.008  0.627     |
20        |  0.012  0.153  0.167  0.008     |
21        |  0.167  0.258  0.825  0.502     |
22        |                                   |
23        |  1.370  0.762  1.000  1.137     | 4.269
24        |                                   |TOTAL SS
25        |ANALYSIS OF VARIANCE TABLE        |
26        |                                   |
27        |      SS      df    MSS   F RATIO|
28   -------|--------------------------------|
29   BET MAC| 3.252     2   1.626  30.326   |
30   BET MEN| 0.696     3   0.232   4.326   |
31   WITHIN | 0.322     6   0.054           |
32        |                                   |
33   TOTAL  | 4.269    11                    |
```

Table 13.4 Formulae for 2 way Analysis of Variance

Source	SS	df	MSS
Between **Rows** Factor	$c\sum(\text{Xibar}-G/N)^2$	r - 1	MSS1
Between **Columns** Factor	$r\sum(\text{Xjbar}-G/N)^2$	c - 1	MSS2
Within Factors	$\sum\sum(\text{X}-\text{Xibar}-\text{Xjbar}+G/N)$	(r-1)(c-1)	MSS3
Total	$\sum\sum(\text{X}-G/N)^2$	rc - 1	

where:

Xibar are the **r** row means, each comprised of **c** values

Xjbar are the **c** column means, each comprised of **r** values

$G = \sum\sum X$ is the grand total

N = rc is the total number of observations

and **MSS1**, **MSS2** and **MSS3** are the ratios of the sums of squares, **SS**, in each row to the corresponding numbers of degrees of freedom, **df**. We form two **F** ratios, **MSS1/MSS3** and **MSS2/MSS3** to test our hypotheses about factor effects.

Table 13.3 shows the data and calculations for a two way analysis of variance. The layout of the spreadsheet is similar to that for a one way analysis of variance but now we have totals and means calculated for columns as well as for rows. The various row means form a column which is titled **Xibar**, and the column means form a row, which is labelled **Xjbar**. Sums of squares are calculated for the two factors, machines in the rows and men in the columns, whose effects we are considering. A total sum of squares is calculated also, but the within sum of squares has been obtained as the difference of the first two sums of squares from the third. The **F** tests yield values of 30.326 with 2 and 6 degrees of freedom, and 4.326 with 3 and 6 degrees of freedom. After File Combining the spreadsheet from table 8.7, enter these values as appropriate. You will find that they are significant at the 0.001 and 0.06 levels respectively. It seems that the different machines do affect output, but that there is some doubt about whether the men do so.

13.4 Summary

Spreadsheets provide a particularly appropriate means of performing analysis of variance calculations. You can see exactly how each sum of squares is formed and how the analysis of variance table is built up.

The facility to copy formulae to rectangular ranges means that you need enter only once each instruction to calculate a set of squared deviations.

14
Time Series Analysis

14.1 Introduction

Techniques of **time series analysis** are needed where some quantitative data have been collected regularly over a period. This may be daily (such as closing stock prices), weekly, monthly (such as government statistics), quarterly, annually (such as company results) or even longer. The usual reason for doing this is to forecast future values in the series. For instance, many businesses seek to predict their future sales on the basis of their past ones. This means that orders for supplies and store inventories can be optimised for product demand. Time series analysis involves only simple calculations carried out in sequence, and is an application for which spreadsheets are especially suitable. Other reasons for time series analysis include making comparisons of current progress with that in the past, or with other firms or countries.

14.2 Graphs

Enter the data from table 14.1 into a worksheet. The quickest way to enter the periods is using the Data Fill option.

The first thing to do with any data set is, of course, to explore the data by graphing it. We need to look for general patterns in the movement of the data. Time is conventionally shown on the X axis, whilst the data themselves are displayed on the Y axis, that is as a 1-2-3 A range. The default graph type, Line, is ideal for time series, although an XY plot can also be used.

Table 14.1 Simple Time Series

	A	B
1	Period	Data
2	1	206
3	2	245
4	3	185
5	4	169
6	5	162
7	6	177
8	7	207
9	8	216
10	9	193
11	10	230
12	11	212
13	12	192
14	13	162
15	14	189
16	15	244
17	16	209
18	17	207
19	18	211
20	19	210
21	20	173
22	21	194
23	22	234
24	23	156
25	24	206

Notice that the data vary considerably over time but that no obvious trend or cycle is present. Such a pattern is referred to as a **stationary** or **horizontal** series. The obvious model to apply is:

$$\text{Data} = \text{Mean} + \text{Residuals} \qquad [14.1]$$

A variation on this model applies when the graph shows an overall drift or **trend** in the series. We would then use:

$$\text{Data} = \text{Trend} + \text{Residuals} \qquad [14.2]$$

Trends are persistent overall long-term tendencies, typically caused by developments in technology or changes in wealth over several years. They can be analysed in various ways. Here, moving means and exponential smoothing will be used to estimate them, but other techniques, including regression (see chapter 12), can also be used.

Time Series Line Graph

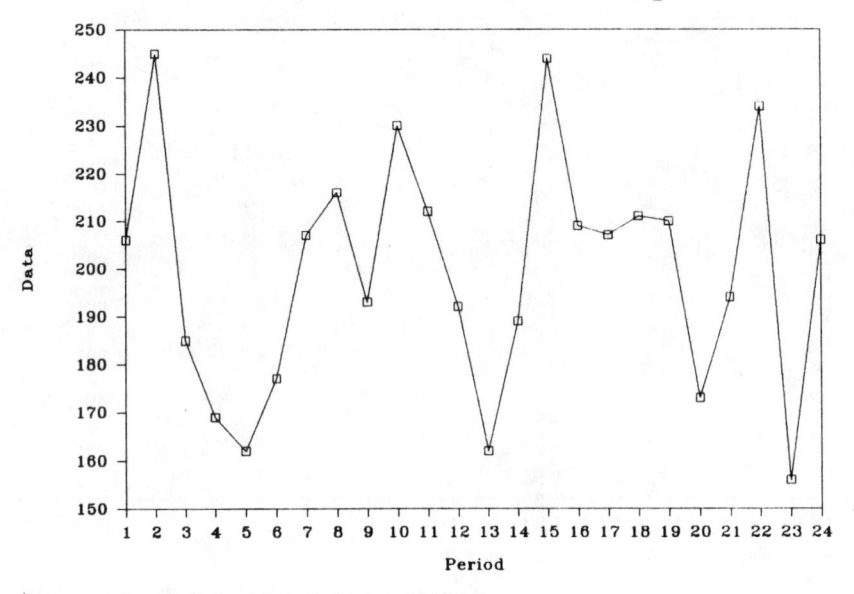

Fig. 14.1 Time Series Line Graph

14.3 Moving Means

Moving means or **averages** are a method of smoothing a time series by averaging successive groups of values. The simplest are three-period moving means in which the observations are averaged in overlapping groups of three. In our example, move to cell C3 and evaluate the average of the first three data values. This can then be copied down the rest of the column. Notice that we have begun alongside the second datum value (B3) since we are averaging that with its predecessor and successor. Notice also that the averages must stop opposite the penultimate datum value (B24) since we need its successor as well. The averages we calculate give estimates of the **trend values** corresponding to the data values in the adjacent column.

In general, if we have **n** data values in total and we are calculating moving means composed of m data values, then we will generate **n-m+1** moving means. So our 24 data values have given us 22 three-period moving means. The smoothing effect of the moving means can be seen by setting up their column as a graph B range and Viewing the graph shown in fig. 14.2 where these values are superimposed on the original data A range.

Three-Period Moving Mean Line Graph

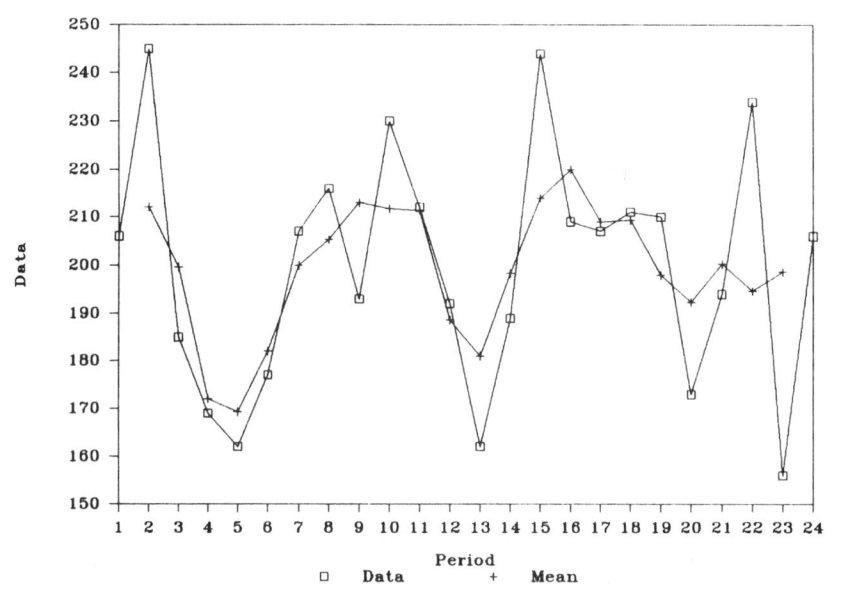

Fig. 14.2 Three-Period Moving Mean Line Graph

A five-period moving mean can be set up in cell D4, copied down and then displayed as a C range. Notice that this curve is even smoother than the previous one, but that it is necessarily shorter too. So, as the number of values included in each moving mean increases so does the smoothing effect, but the number of means also decreases. At the limit, when the number of values in the moving mean is equal to the total number of data values we have just one overall or grand mean! If the series is completely random, such that each value bears no relationship to its predecessor, then this one value is the best forecast we can make about the future of the series.

Where there is regular **periodicity** in the data, such as a seasonal influence, the number of time periods included in the moving mean should be that which comprises a complete cycle. So for quarterly data a four period moving mean is needed, for monthly data a twelve period one. Since these are even numbers, these moving means require centring so that they correspond to actual observations, and only **n-m** moving means are then formed.

The moving mean method has the disadvantage that it gives equal weight to all its constituent values even though they are of differing ages and, possibly, relevance.

14.4 Exponential Smoothing

This method avoids the disadvantage of the moving mean approach by weighting according to the following formula:

$$\text{Forecast}(t+1) = a*\text{Data}(t)+(1-a)*\text{Forecast}(t) \qquad [14.3]$$

In other words, the forecast for the next time interval is equal to some smoothing parameter, **a**, multiplied by the data value for the previous time period, added to one minus the smoothing factor multiplied by the forecast value for the previous time period. The smoothing factor must lie between 0 and 1, is commonly between 0.05 and 0.3, and is frequently set to 0.2.

Continuing with the spreadsheet you have set up, in cell E1 enter a smoothing value of 0.1. Range Name Create this as **Alpha**. Exponential smoothing requires an initial forecast as a starting point, so copy the first datum value across into cell E2. In cell E3 recast the right-hand side of [14.3] as a 1-2-3 formula, that is:

$$\text{+\$Alpha*B2+(1-\$Alpha)*E2} \qquad [14.4]$$

Notice that this formula consists of both absolute and relative addresses, the former being the smoothing factor and the latter being the data and forecast. This formula can then be copied down the rest of the column. In fact it can be copied down into one more cell than the data, giving a forecast for one period beyond the data.

In the Graph menu, Reset the B and C ranges and then make the exponentially smoothed forecasts into a B range.

You can try changing the smoothing factor and seeing its effect on the graph. At the extremes, a value of 0 will give a horizontal line at the starting value, while a value of 1 will give a curve indentical to the data but lagging one period behind it. Thus values close to zero give most smoothing, and values close to unity give least.

Data & Exponentially Smoothed Forecast

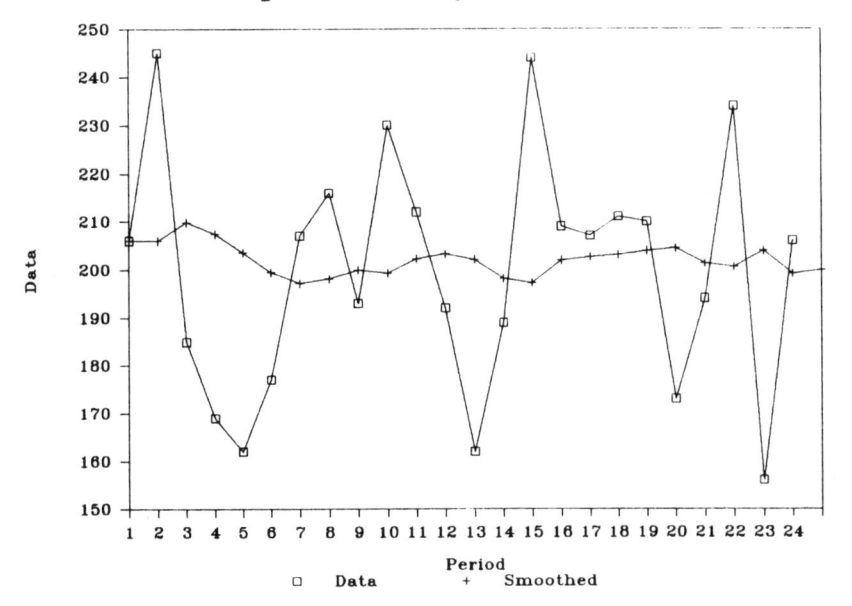

Fig. 14.3 Data and Exponentially Smoothed Forecasts

14.5 Seasonal Analysis

Many time series show a **regular cyclical pattern**. Two of the most important are the daily cycle of day and night caused by the earth's rotation, and the annual cycle of the four seasons caused by the earth orbiting the sun on a tilt. The former gives rise to a distinctive cycle in electricity demand, while the latter causes variations in the prices of food and other goods. Other, less regular, patterns include the business cycle which varies from 2 to 10 years in duration.

After deseasonalising a time series the trend may be a curve, which may be transformed into a straight line by applying various mathematical functions such as logarithms and square roots (see Chapter 16). The resulting straight line can then be extrapolated, and the inverse of the function then applied to give a forecast.

Enter the data from table 14.2 into a worksheet.

Table 14.2 UK Air Passenger Movements

```
      A      B    C     D      E      F
1    UK PASSENGER MOVEMENTS BY AIR
2
3    YEAR  QTR  DATA  MOVE   TREND   Tt-
4    Y     Q    D     MEAN   T       Tt-1
5    1982  3    7623
6          4    4374
7          1    3924
8          2    6018
9    1983  3    8100
10         4    4723
11         1    4172
12         2    6804
13   1984  3    8747
14         4    5022
15         1    4537
16         2    6848
17   1985  3    9119
18         4    5389
19         1    4992
20         2    7364
```

Set up the quarters as a Line Graph X range and the data as an A range and then View the result. Notice that the quarters are displayed equally spaced along the horizontal axis; you would not use an XY graph for this data. The data can be seen to have an annual cycle imposed on a longer term trend. This data set can be analysed by several methods, of which the additive and multiplicative models are the best known. These models will be applied to the data for years 1982-84, and predictions made from them for 1985.

14.6 The Additive Model

The additive model to fit such data is

$$\text{Data} = \text{Trend} + \text{Seasonal} + \text{Random} \qquad [14.5]$$

The first step is to estimate the trend. This can then be subtracted from the data to give the seasonal component plus random fluctuations. We shall use moving means as our trend estimates, so in D6 calculate a moving mean of C5 to C8 and copy this down, as shown in table 14.3. Notice that this is a four-period average which really belongs between the second and third periods at 2.5! We can resolve this problem by evaluating, in E7, a centred moving mean composed of the average of D6 and D7. That is, we average the moving means for 1983 Q0.5 and Q1.5 to give a centred moving mean at Q1. Our 12 datum values give us 9 ordinary moving means but only 8 centred moving means.

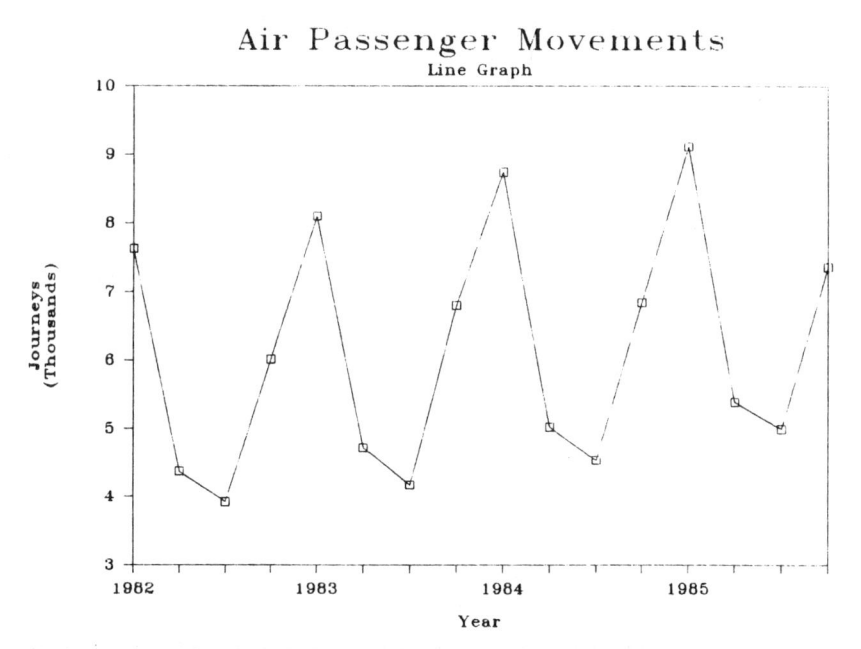

Fig. 14.4 Air Passenger Movements Line Graph

In cell G7 we can compute the data minus the trend as +C7-E7 and copy it down to G14. In H7 we can find the seasonal effect by averaging the differences for the same period of the year in different years. We average first the third quarter values, that is (G7+G11)/2. This can be copied down to H10 to give all four seasonal effects. In H13 evaluate the mean of the seasonal effects. For the seasonal effects to be centred on the trend this mean should be zero. Although not shown in our spreadsheet, a non-zero mean could be subtracted from each of the seasonal factors to ensure they do have zero mean.

The data can then be deseasonalised by evaluating +C7-H7 in cell I7. This can be copied down only as far as cell I10. The formulae for the rest of this column, both above and below the four values already generated, must subtract the appropriate seasonal effect from the original datum values. For example, the formula in cell I5 is +C5-H9, in I11 it is +C11-

H7 and in I15, +C15-H7. These can be copied down 2, 4 and 2 cells respectively to complete the column of deseasonalised data. Reset the Graph to show all the original data and the adjusted data (Fig. 14.5).

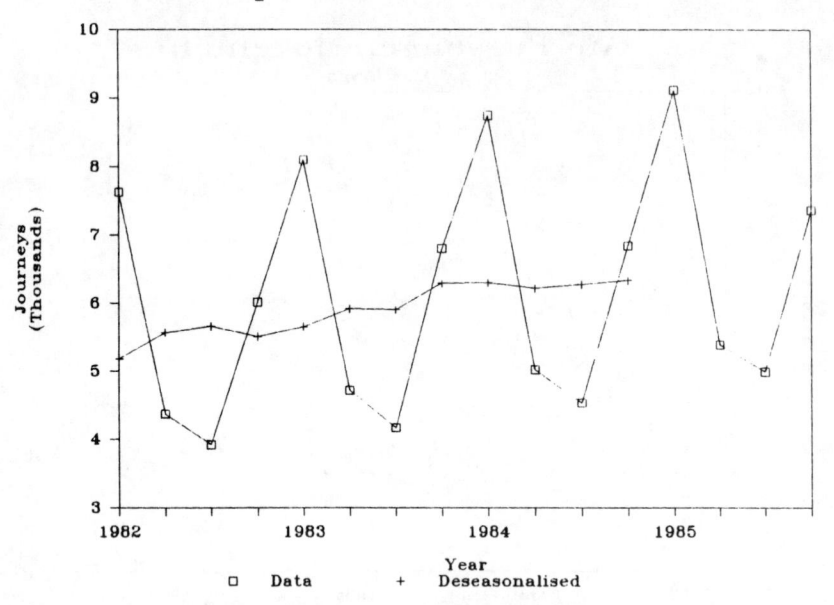

Fig. 14.5 Line Graph with Deseasonalised Values

In order to extrapolate the moving means we need to find the trend differences:

Trend difference(t) = Trend(t) - Trend(t-1)   [14.6]

So in cell F8 evaluate +E8-E7 and then copy this down to F14. In F17 we can find the average trend difference. Assuming that the trend is linear, this allows us to find further trends according to the formula:

Trend(t+1)=Trend(t)+Average trend difference   [14.7]

To do this, move to E15, evaluate +E14+$F$17 and then copy this down to E20.

We can now apply the seasonal effects to the extrapolated trends to give a forecast which we can compare with the actual observations (Fig. 14.6).

In H15 compute +E15+H7 and copy this down to H18 and then to H20. Reset the Graph and set the X range as B15..B20, the A range as C15..C20 and the B range as H15..H20.

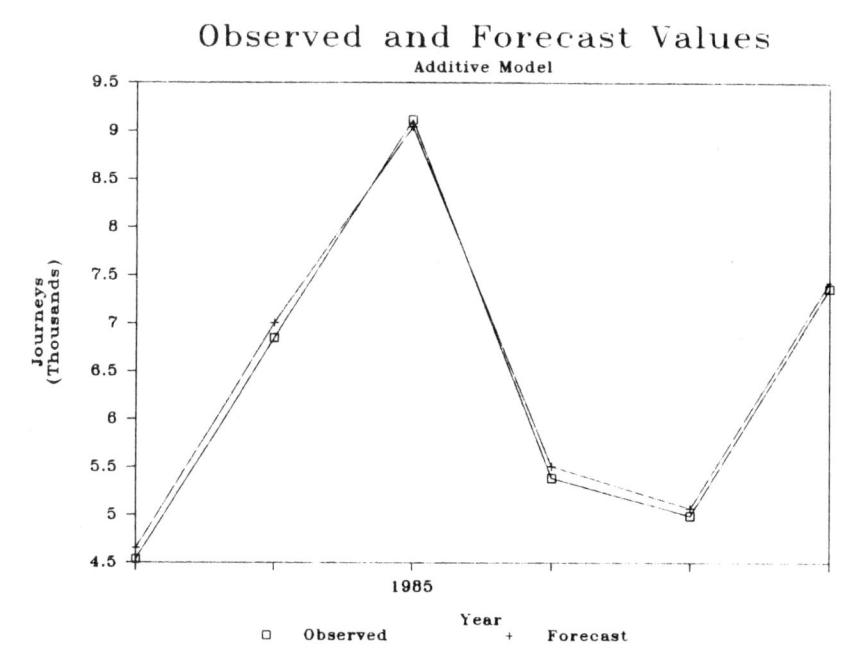

Fig. 14.6 Observed and Forecast Values, Additive Model

14.7 The Multiplicative Model

The alternative, multiplicative, model to fit such data is

Data = Trend * Seasonal * Random [14.8]

In cell J7 we can compute the data divided by the trend as +C7/E7 and then copy it down to J14. In K7 we can find the seasonal effect by averaging the third period differences, that is (J7+J11)/2. This can be copied down to K10 to give all four seasonal effects.

In K13 find the mean of the seasonal effects. Since this is unity, the seasonal factors are balanced around the trend. If they were not, an appropriate correction factor could be applied.

We can now apply the multiplicative seasonal effects to the extrapolated trends to give a forecast which we can compare with the actual observations (Fig. 14.7). In K15 compute +E15*K7 and copy this down appropriately to K20. Reset the Graph and set the X range as B15..B20, the A range as C15..C20, the B range as H15..H20 and the C range as K15..K20.

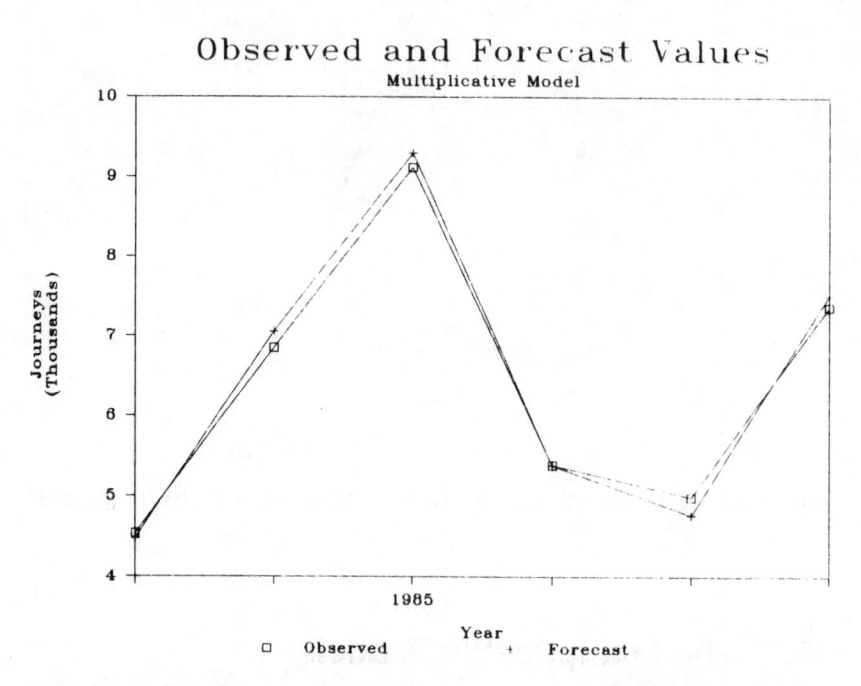

Fig. 14.7 Observed/Forecast Values, Multiplicative Model

The data can then be deseasonalised by evaluating +C7/K7 in cell L7. This can be copied down only to cell L10. The formulae for the rest of this column, both above and below the four values already generated, must divide the original datum values by the appropriate seasonal effect. For example the formula in cell L5 is +C5/K9, in L11 it is +C11/K7 and in L15, +C15/K7. These can be copied down 2, 4 and 2 cells respectively to complete the column of deseasonalised data. The final spreadsheet is shown in table 14.3.

Table 14.3 Additive and Multiplicative Seasonal Analysis

	C	D	E	F	G	H	I	J	K	L
1	UK PASSENGER MOVEMENTS BY AIR								
2						ADDITIVE MODEL:		MULTIPLICATIVE:	
3	DATA	MOVE	TREND	Tt-	DIFF.	EFFECT	ADJ.	RATIO	EFFECT	ADJ.
4	D	MEAN	T	Tt-1	D-T	S1	DATA	D/T	S2	DATA
5	7623						5177			5408
6	4374	5485					5569			5446
7	3924	5604	5544		-1620	-1740	5664	0.71	0.70	5608
8	6018	5691	5648	103	370	513	5505	1.07	1.09	5541
9	8100	5753	5722	75	2378	2446	5654	1.42	1.41	5746
10	4723	5950	5852	129	-1129	-1195	5918	0.81	0.80	5880
11	4172	6112	6031	179	-1859		5912	0.69		5962
12	6804	6186	6149	118	655	MEAN:	6291	1.11	MEAN:	6265
13	8747	6278	6232	83	2515	6.234	6301	1.40	1.000	6205
14	5022	6289	6283	51	-1261		6217	0.80		6252
15	4537		6389			4649	6277		4471	6484
16	6848		6494	MEAN:		7007	6335		7053	6305
17	9119		6600	105.		9046			9302	
18	5389		6705			5510			5386	
19	4992		6811			5071			4766	
20	7364		6916			7429			7511	
21	DATA		TREND		ADDITIVE FORECAST				MULTIPLICATIVE	

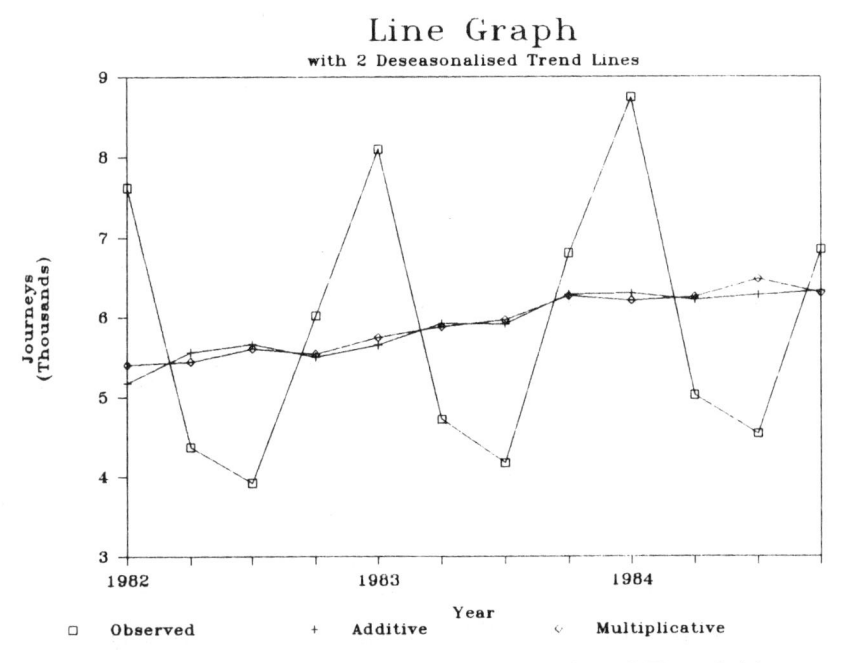

Fig. 14.8 Line Graph with 2 Deseasonalised Trend Lines

165

Reset the Graph to the original data and the two adjusted sets to give fig. 14.8.

14.8 Summary

Notice that the two models have given essentially the same adjusted data in this case. The additive model is more appropriate where the amplitude of the cycles is roughly constant, while the multiplicative model is more appropriate when the amplitude varies in relation to the trend itself.

If you average the seasonal effects in columns H and K they may not quite come to the expected values of zero or one respectively. Correction factors can be used to compensate for this. These can be simply equal weights or proportional values applied to the data.

The 'best' deseasonalising method can be chosen by selecting the lowest standard deviation of the various adjusted data sets. This can, of course, be easily calculated with a spreadsheet.

15
Financial Calculations

- @IRR
- @NPV
- Movement in large spreadsheets
- @PMT
- @PV
- @FV

15.1 Introduction

In this book we have sought to show that the spreadsheet is a general purpose tool that can profitably be used whenever numbers need to be manipulated. However, spreadsheets were initially designed as special purpose vehicles for financial calculations and this is still their prime use in business. This chapter covers compound interest and the five financial functions built into 1-2-3.

15.2 Compound Interest

If a capital sum of money is deposited with a bank it will earn interest over a period of time. If the interest is left to accumulate with the capital then it is called **compound interest**. For example £100 deposited at a rate of 10% interest payable annually will earn interest of £10 in the first year. Adding this to the initial capital gives a capital sum of £110 at the end of year 0. During the next year, interest of £11 will be earned on this sum so that at the end of year 1 the capital will be worth £121. Let us set up the worksheet shown in table 15.1 to calculate compound interest for 5 years.

Table 15.1 Calculation of Compound Interest

```
        A         B         C         D
1  COMPOUND INTEREST
2  Year:     Rate:     Capital:  Interest:
3           0         10%       100.00    10.00
4           1         10%       110.00    11.00
5           2         10%       121.00    12.10
6           3         10%       133.10    13.31
7           4         10%       146.41    14.64
8           5         10%       161.05    16.11
9
10 Initial capital        Rate          t
11      100                10%          5
12
13 Initial Capital*(1+Rate)^t
14   161.05
```

The years can be entered with Data Fill. The column headed **Rate** contains interest rates, entered in such a way that they will all alter if the top one is changed. You should begin by entering 10/100 in B3, and bring the rate down using a formula by entering +B3 in cell B4. Now display cells B3 and B4 as percentages using Range Format Percent 0. The initial capital sum of £100 is entered in C3 whilst the interest earned in year 0 is calculated in cell D3. Since for any given time period

Interest = Rate * Capital [15.1]

you should enter the formula

+C3*B3

to multiply the initial capital by the interest rate. For any time period **t** we have

Capital(t+1) = Capital(t) + Interest(t) [15.2]

The capital at the beginning of year 1, then, is the capital at the beginning of year 0 plus the interest earned in that year. To calculate this, enter

+C3+D3

in cell C4. Copy the **Interest** formula from D3 to D4. Then simultaneously copy the **Rate**, **Capital** and **Interest** formulae for year 1 to the bottom of the table. Notice that the percent format for the interest rate is copied also.

The compound or exponential growth can be inspected with a graph. We show this in fig. 16.2.

The capital accumulated at any stage can be calculated by the following formula:

Capital = Initial Capital * (1 + Rate)^t [15.3]

where **t** is the current time period. Appropriately labelled values for these expressions are displayed at the foot of table 15.1. Use the formulae +C3 and +B3 to bring down the initial **Capital** and **Rate** values so that the whole table will recalculate if either value is changed.

Range Name Labels Down is used to assign the labels to the values as names. This allows us to calculate the final capital by entering a value of 5 for **t**, and a formula which is the right hand side of 15.3, preceded by a [+] sign. A value of 161.05 is calculated, which is the same as that shown in the last row of the table.

15.3 Net Present Value

- @IRR
- @NPV

If we wish to provide money now which will grow to a particular capital sum at some specified future date we can investigate the initial capital required using the spreadsheet shown in table 15.1. Simply enter the appropriate value of **t**, and then try enterring alternative values of **Initial Capital** in the cell C3 until the desired final **Capital** value is displayed.

A direct expression for initial capital is obtained by rearranging equation 15.3. This gives:

Initial Capital = Capital * (1/(1+Rate)^t) [15.4]

When the formula is used in this form the interest rate is often called the **discount rate**, and the initial capital is called the **discounted value**, or **present value** of the final one. Notice, however, that in 1-2-3 the term present value has a rather different meaning which we shall see in section 15.4 below.

Here we are concerned with **Net Present Value**, **NPV**, which is the aggregate of a series of discounted values from different future time periods. The most common application of **NPV** involves comparing an initial capital expense against some projected future income receipts.

Take, for example, an initial capital **cost** of £1000 which is expected to generate **returns** of £500 in the first year, £600 in the second and £700 in the third and final year of the project. The interest rate, now renamed the **discount rate**, is chosen to be 10%. This can be entered as 10/100, brought down as a formula and then displayed with Range Format Percent 0. Notice that the capital cost is entered as a negative cash flow.

Set this up in the following worksheet:

Table 15.2 Data for NPV Calculation

	A	B	C	D	E	F
1	INVESTMENT	ANALYSIS	Cost/	Discount	Discounted	NPV
2	Year:	Rate:	Return:	Factor:	Value:	to Date:
3	0	10%	-1000			
4	1	10%	500			
5	2	10%	600			
6	3	10%	700			
7						
8	IRR =				NPV =	

The **discount factor** is derived from formula 15.4 as the multiplicand of **Capital** and is entered in D3 in the form 1/(1+B3)^A3. For the base year, the **discounted value** is therefore +C3*D3 and the **NPV to Date** is the same as the discounted value (+E3). Copy the **Discount Factor** and **Discounted Value** formulae down one row, then calculate the **NPV to Date** for Year 1 by adding the discounted value for the current year to the previous figure in the **NPV to Date** column. That is, you should enter the formula +F3+E4 in F4. The remaining blank cells in the table can now be completed by simultaneously copying down the formulae in D4..F4.

Table 15.3 NPV Calculation

	A	B	C	D	E	F
1	INVESTMENT	ANALYSIS	Cost/	Discount	Discounted	NPV
2	Year:	Rate:	Return:	Factor:	Value:	to Date:
3	0	10%	-1000	1.000000	-1000.00	-1000.00
4	1	10%	500	0.909090	454.55	-545.45
5	2	10%	600	0.826446	495.87	-49.59
6	3	10%	700	0.751314	525.92	476.33
7						
8	IRR =		34%		NPV =	1476.33

This shows that the overall **NPV** for the project is £476.33. A positive **NPV** indicates that the project is profitable.

An alternative evaluation method is to compute the **Internal Rate of Return**, **IRR**. This is defined as the discount rate that makes the **NPV** of the series of cash flows zero. You can use the spreadsheet shown in table 15.3 to find the **IRR** by a trial and error procedure. A larger discount rate will lower the overall **NPV**, so try entering different values at the top of the **Rate** column until the last value in the **NPV to date** column is 0. A better method is to use a Data Table.

Alternatively, you can use the built-in **IRR** function. It has the general form:

@IRR(**Guess,Range**)

where **Guess** is a starting value for the calculations, which should be a number between 0 and 1, and **Range** covers the cash flows. In our example @IRR(1,C3..C6) gives the value 34% Check that this is the value you obtained by your trial and error procedure. The **IRR** is the discount rate at which a project just breaks even. There are sometimes difficulties in using it as an investment criterion, but for this project we can say that at interest rates lower than 34% the project is profitable and at higher rates it makes a loss.

Finally we can use the built-in function @NPV to check our earlier result and to plot a graph of **NPV** for various discount rates. This uses the formula:

$$NPV = \sum_{t=1}^{n} \{ V_t / (1+\text{Rate})^t \}$$

where V_t stands for the costs or returns at time **t** and **Rate** is the discount rate. Notice that the summation starts from time period 1, not zero. The function has the general form:

@NPV(**Rate,Range**)

where **Range** consists of the costs or returns without the value in year 0. In the spreadsheet shown in table 15.3, the formula @NPV(B6,C4..C6) generates the answer £1476.33, but taking the initial capital expense into account
(-£1000), then the true NPV is £476.33 again.

Table 15.4 has the same **Cost/Return** values as table 15.3 and uses the @NPV function to build up a table of NPV's at different discount rates. The results of applying @NPV at the various discount rates are shown in the **NPV of Returns** column. The final column, **Overall NPV**, is derived by adding the already negative cost figure to these values and thus subtracting it.

Table 15.4 NPV at Various Discount Rates

```
         A           B          C         D          E
1   INVESTMENT ANALYSIS - ALTERNATIVE DISCOUNT RATES
2                Cost/    Discount     NPV of    Overall
3       Year:   Return:    Rates:    Returns:      NPV:
4          0    -1000        5%      1625.09     625.09
5          1      500       10%      1476.33     476.33
6          2      600       15%      1348.73     348.73
7          3      700       20%      1238.43     238.43
8                           25%      1142.40     142.40
9                           30%      1058.26      58.26
10                          35%       984.10     -15.90
11                          40%       918.37     -81.63
```

NPV at Various Discount Rates

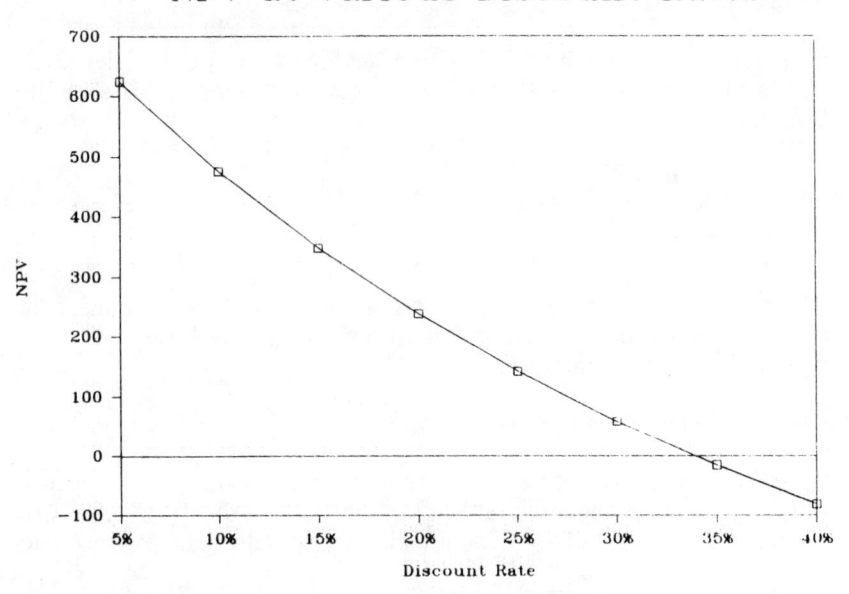

Fig. 15.1 NPV at Various Discount Rates

Table 15.4 shows that discount rates of 35% and 40% yield negative **NPV**'s, as we would expect from our **IRR** calculations. Fig. 15.1 is an XY graph of the overall **NPV** values plotted against the discount rates. Notice how **NPV** falls as the discount rate rises, and that it becomes negative when the discount rate is 0.34, or 34%.

15.4 Regular payments

- **[HOME], [END], [PGUP], [PGDN]**
- **Function key [5/GOTO]**
- **@PMT**
- **@PV**
- **@FV**

When a capital sum (a principal) is borrowed, such as a mortgage on a house or a loan for a car, then it can be paid off as a series of payments at some interest rate over some term.

Suppose a £50,000 mortgage is taken out at an interest rate of 14.5% over 25 years. We shall set up a spreadsheet, parts of which are shown in tables 15.5 and 15.6, to illustrate the repayment process, and we shall be able to find the monthly repayment required by a trial and error procedure.

The spreadsheet we shall set up in this section is much larger than those we have used previously. To move about in it you will find the special movement keys useful. **[HOME]** moves you to cell A1; **[END]** followed by **[UP]**, **[DOWN]**, **[LEFT]** or **[RIGHT]** moves you to the end of the section in the direction you have specified. You can move one screenful at a time using **[PGUP]** and **[PGDN]**. Another alternative is to use function key **[5/GOTO]** and enter the cell address or range name to which you wish to move.

Table 15.5 Regular Repayments of £650

	A	B	C	D	E	F
1	REGULAR	RE-PAYMENTS	Out-		After	
2	Month:	Rate:	standing	Payment:	Payment:	Interest:
3	0	1.21%	50000.00	0	50000.00	604.1667
4	1	1.21%	50604.17	650	49954.17	603.6128
5	2	1.21%	50557.78	650	49907.78	603.0523
6	3	1.21%	50510.83	650	49860.83	602.4851
7	4	1.21%	50463.32	650	49813.32	601.9109
8	5	1.21%	50415.23	650	49765.23	601.3298
:						
223	220	1.21%	1161.81	650	511.81	6.1844
224	221	1.21%	517.99	650	-132.01	-1.5951
:						

Use Data Fill to enter the months for the 25-year repayment period. For the range enter [END][DOWN] to indicate all the rows, and for the stop value enter the formula 25*12 to give the value of the last month as 300. This is shown in table 15.6 but not in table 15.5. Notice that both tables show only parts of the spreadsheet. When we have set up the rows for months 0, 1 and 2 we shall want to copy the month 2 formulae to the bottom of the table. To make it easy to define the end point of the large Copy To range, repeat the Data Fill process you have just undertaken, filling values which will later be overwritten into the **Rate** column.

Now enter the monthly interest **Rate**, which is the annual rate of 14.5% divided by 12, for month 0. This will overwrite one of your filled values. Enter the amount **Outstanding** in month 0, which is £50,000, the amount borrowed. No repayment is made during month 0, so enter a **Payment** value of 0. The **After Payment** column calculates the difference between the amount outstanding at the beginning of the month and the payment made, so for month 0 we obtain the value 50,000. **Interest** is incurred on this amount during the month, and is therefore calculated by multiplying the amount outstanding after payment by the interest rate. This gives the interest incurred during month 0 as £604.1667. Adding this to the amount outstanding after payment for month 0 gives the value outstanding at the beginning of month 1, so a formula to add the last 2 column values for month 0 should be entered in the **Outstanding** column for month 1. Enter a trial value for the monthly payment in the month 1 **Payment** cell. Table 15.5 shows a value of £650. Use a formula to bring the interest **Rate** down to month 1, and Copy the formulae in the last 2 columns to the month 1 row. The month 2 **Payment** must be obtained by a formula to enable recalculation to take place when changes are made, but the other month 2 values can be obtained by copying.

You can now copy your month 2 formulae to the bottom of the table. Because you earlier filled values to the bottom of the range to which you now wish to copy, you can define the Copy To range by typing the [.], stretching the range to cover the formulae, then pressing [END] followed by [DOWN]. This stretches the range to include the last filled value, so that pressing [RETURN] completes the table.

Move the cell pointer so you can see the values for month 221. There is £517.99 outstanding at the beginning of the month, so that a further monthly payment of £650 more than pays off the debt. With this monthly payment the mortgage would be paid of in 221 months instead of in 300.

Try entering different values for the first monthly payment. You should find that a payment of £621.08 almost exactly pays off the mortgage in the term specified. Table 15.6 shows a balance of £4.31 outstanding after the 300th payment is made. The formula for directly calculating the repayment is:

Payment = Principal*Rate/{1-(1+Rate)^-Term}   [15.5] and 1-2-3 has a built-in function, @PMT, which evaluates it. The function takes the general form:

@PMT(**Principal**,**Rate**,**Term**)

Table 15.6 Regular Repayments of £621.08

	A	B	C	D	E	F
1	REGULAR RE-PAYMENT		Out-		After	
2	Month:	Rate:	standing	Payment:	Payment:	Interest:
3	0	1.21%	50000.00	0	50000.00	604.1667
4	1	1.21%	50604.17	621.08	49983.09	603.9623
5	2	1.21%	50587.05	621.08	49965.97	603.7555
6	3	1.21%	50569.72	621.08	49948.64	603.5461
7	4	1.21%	50552.19	621.08	49931.11	603.3343
8	5	1.21%	50534.44	621.08	49913.36	603.1198
:						
302	299	1.21%	1239.00	621.08	617.92	7.4666
303	300	1.21%	625.39	621.08	4.31	0.0521

To check the calculation shown in table 15.6, enter

@PMT(50000,14.5%/12,25*12)

which gives an answer of £621.0815 as the monthly repayment. Notice that the interest rate is divided by 12 and the term is multiplied by 12 to change from years to months.

Table 15.7 Regular Payments from a Capital Sum

	A	B	C	D	E	F
1	REGULAR PAYMENTS		Out-		After	
2	Month:	Rate:	standing	Payment:	Payment:	Interest:
3	0	1.00%	700.00	0	700.00	7.0000
4	1	1.00%	707.00	60	647.00	6.4700
5	2	1.00%	653.47	60	593.47	5.9347
6	3	1.00%	599.40	60	539.40	5.3940
7	4	1.00%	544.80	60	484.80	4.8480
8	5	1.00%	489.65	60	429.65	4.2965
9	6	1.00%	433.94	60	373.94	3.7394
10	7	1.00%	377.68	60	317.68	3.1768
11	8	1.00%	320.86	60	260.86	2.6086
12	9	1.00%	263.47	60	203.47	2.0347
13	10	1.00%	205.50	60	145.50	1.4550
14	11	1.00%	146.96	60	86.96	0.8696
15	12	1.00%	87.83	60	27.83	0.2783

If the opposite situation applies, where we want a regular income from a capital sum, we can use the spreadsheet we have set up in a different way. Enter as the month 1 **Payment** the monthly income you require. Make a guess at the initial capital sum required and enter this value in the

Outstanding column for month 0. Enter successive guesses until the last figure shown in the **After Payment** column is approximately 0.

If you wish to provide a monthly payment of £60 over a year at an interest rate of 12% p.a., enter £60 for the **Payment** and guess £700 as the initial **Outstanding** value. Table 15.7 shows that there will be £27.83 outstanding after the twelfth payment is made. Try entering initial **Outstanding** values of rather less than £700, and you will find that £675.30 is the capital sum required.

Formula 15.5 can be arranged to give this directly:

    Principal = Payment*(1-(1+Rate)^-Term)/Rate   [15.6]

1-2-3 has a built-in function, Present Value, to evaluate this formula:

    @PV(**Payment,Rate,Term**)

Check your previous result by entering

    @PV(60,12%/12,12)

which again gives a value of £675.30.

Note that Present Value assumes that the number of payments and the compounding period are the same and that the payments are made at the end of each period.

Table 15.8 Regular Payments to form a Capital Sum

	A	B	C	D	E	F
1	REGULAR PAYMENTS		Out-			
2	Month:	Rate:	standing	Payment:	Sum:	Interest:
3	0	1.00%	0.00	0	0.00	0.0000
4	1	1.00%	0.00	60	60.00	0.6000
5	2	1.00%	60.60	60	120.60	1.2060
6	3	1.00%	121.81	60	181.81	1.8181
7	4	1.00%	183.62	60	243.62	2.4362
8	5	1.00%	246.06	60	306.06	3.0606
9	6	1.00%	309.12	60	369.12	3.6912
10	7	1.00%	372.81	60	432.81	4.3281
11	8	1.00%	437.14	60	497.14	4.9714
12	9	1.00%	502.11	60	562.11	5.6211
13	10	1.00%	567.73	60	627.73	6.2773
14	11	1.00%	634.01	60	694.01	6.9401
15	12	1.00%	700.95	60	760.95	7.6095

We may wish to know the final value of a regular series of payments, for example monthly savings. Table 15.8 is a slightly modified version of our previous spreadsheet which will calculate this. This time we want to **add** the **Payment** to the sum **Outstanding**, so we alter the title of the penultimate column and the formulae in it. Suppose we deposit £60 per month at 12% p.a. Starting from an outstanding amount of 0, after 60 monthly payments we have a final sum of £760.95. The formula to calculate this directly is:

Final Capital = Payment*(((1+Rate)^Term)-1)/Rate [15.7]

The value can be found from 1-2-3's Future Value function which takes this general form:

@FV(**Payment,Rate,Term**)

Check your example to see that your savings will be worth @FV(60,12%/12,12) = £760.95.

Note that Future Value assumes that the number of payments and the compounding period are the same, and that the payments are made at the end of each period.

15.5 Summary

This chapter has shown how spreadsheets can be set up to simulate financial processes, giving insight into what it is that is calculated by 1-2-3's financial functions.

16
Transformations

- @EXP
- @LN
- @LOG

16.1 Introduction

Many statistical methods can be applied only when certain assumptions are fulfilled. For example, various hypothesis tests and confidence interval estimates require that the population values be normally distributed. The technique of linear regression which we studied in chapter 12 is applicable only when the relationship between the variables is linear. In such cases, although our original data may not be suitable, it may be possible to **transform** them into new variables which fulfil the prerequisites, so that a particular statistical method can be applied.

In this chapter we shall see three examples of such transformations. The first is used when a variable has a positively skewed distribution, to transform it into a variable which is more nearly normally distributed. The other examples start with non-linear relationships between two variables and show how these can be transformed into linear ones.

A **transformed** variable is simply some particular function of the original variable. If the latter is displayed in one column of your spreadsheet, you can construct the transformed variable in another column simply by entering and copying an appropriate formula. You will also do this if you are working with a model whose specification includes some function of a variable. One example of this is a lag function, which requires values of the variable for a previous year.

16.2 Positively Skew Distributions

- @EXP
- @LN

The distribution shown in fig.16.1 is **positively skew**: that is, it has a long tail to the right. It is an XY graph of various points on a **log-normal** density function. The variable **X** whose density is being plotted is such that if we take its logarithm to base **e**, forming a new variable $Z = \ln(X)$, this new variable **Z** has a normal distribution such as we graphed in chapter 8.

Log-Normal Density Function

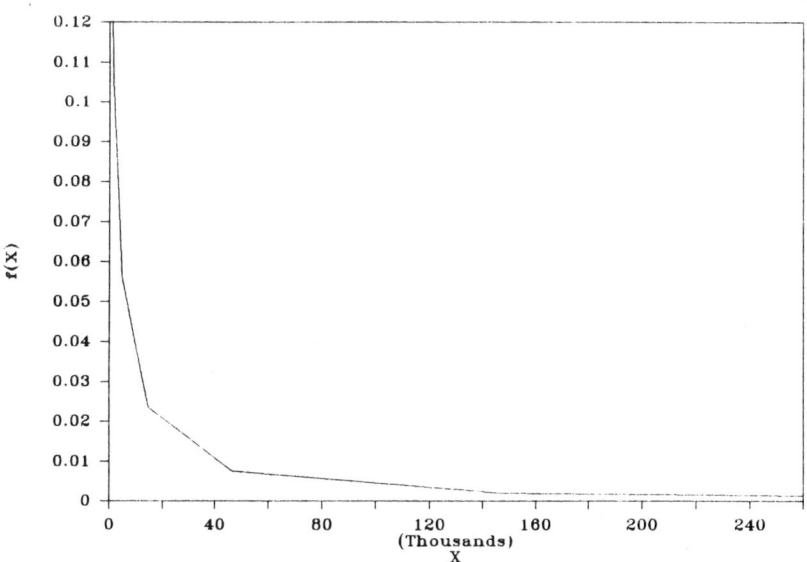

Fig. 16.1 Log-Normal Density Function

Fig. 16.1 is graphed from the spreadsheet shown in table 16.1. The first two columns of this spreadsheet are the last two columns of table 8.4, with the variable re-labelled **Z** instead of **X**. In table 16.1 **Z** is a normal variable and the corresponding values of f(**Z**) give points on its density function. From **Z** we can construct the third column of table 16.1 which contains values of the log-normal variable, **X**. These are formed using the relationship

$$X = e^Z \qquad [16.1]$$

because, taking natural logarithms of both sides, this implies that

$$\ln(X) = Z \qquad [16.2]$$

and **Z** is a normal variable. Each value of **Z** corresponds to a particular value of $X = e^Z$, and so the density functions for corresponding values are the same. That is, f(**X**) = f(**Z**).

Table 16.1 Log-Normal Transformation

	C	D	E
1	Normal	f(X)	Lognormal
2	variable	=	variable X
3	Z	f(Z)	@EXP(Z)
4			
5	-4.2	0.000058	0.01499557
6	-3.0	0.000379	0.04735892
7	-1.9	0.001927	0.14956861
8	-0.7	0.007621	0.47236655
9	0.4	0.023474	1.49182469
10	1.6	0.056312	4.71147018
11	2.7	0.105205	14.8797317
12	3.9	0.153072	46.9930632
13	5.0	0.173453	148.413159
14	6.2	0.153072	468.717386
15	7.3	0.105205	1480.29992
16	8.5	0.056312	4675.07273
17	9.6	0.023474	14764.7815
18	10.8	0.007621	46630.0284
19	11.9	0.001927	147266.625
20	13.1	0.000379	465096.411
21	14.2	0.000058	1468864.18

You should construct table 16.1 from table 8.4 by re-titling your columns and using the function @EXP with the address of the first value of **Z** as its argument to form the first value of **X**. Copying down the formula will complete the table. Check that @LN(**X**), where **X** is any one of your **X** values, referred to by its cell address, gives the corresponding original **Z** value. An XY Graph of **X** against f(**X**), with some manual scaling, will display the log-normal density function shown in fig. 16.1.

Whenever a variable has a distribution similar to that shown in fig. 16.1, the logarithm of that variable will have an approximately normal distribution.

16.3 Exponential Growth

- @LOG

An amount of money growing at compound interest, such as we saw in chapter 15, is one example of **exponential growth**. The time period, **t**, which we Graph as an **X** range, appears as the **exponent** of (1+**Rate**) in formula 15.3 from which the amounts of capital are calculated. Other variables such as population, or, more optimistically, G.N.P. may also grow in the same way.

Table 16.2 Semi-Logarithmic Transformation

	A	B	C	D	E
1	COMPOUND	INTEREST			
2	Year:	Rate:	Capital:	Interest:	Logcap:
3	0	30%	100.00	30.00	2.0000
4	1	30%	130.00	39.00	2.1139
5	2	30%	169.00	50.70	2.2279
6	3	30%	219.70	65.91	2.3418
7	4	30%	285.61	85.68	2.4558
8	5	30%	371.29	111.39	2.5697

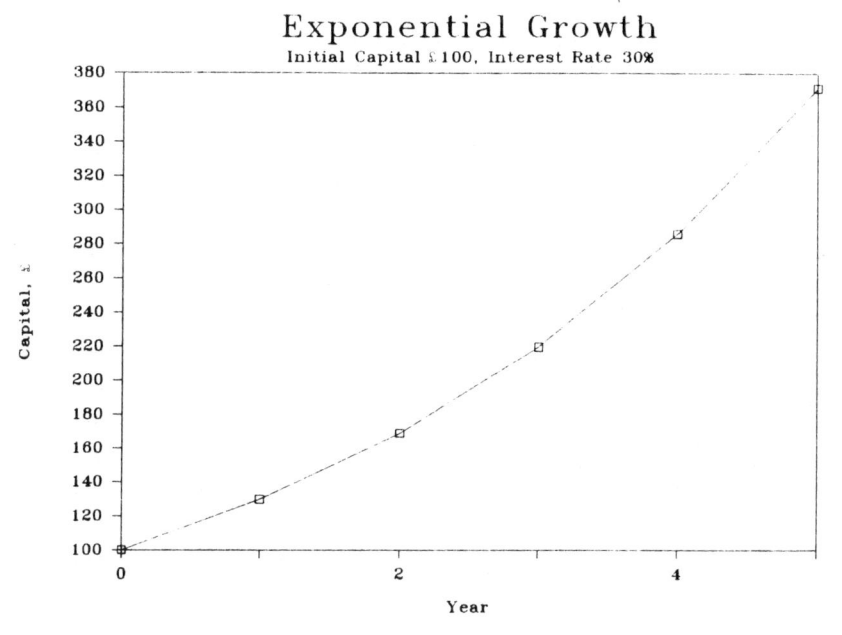

Fig. 16.2 Exponential Growth

Table 16.2 has been constructed from table 15.1. To emphasise the curvature of the growth function the rate of growth, **Rate**, has been increased to 30%. An XY Graph of **Capital** for each **Year** shown in table 16.2 is displayed in fig. 16.2, and we can see that the values form a curve.

The curvilinear relationship shown in fig. 16.2 can be changed into a linear one by what is known as a **semi-logarithmic** transformation. This involves taking the logarithm of **one** of the variables. Table 16.2 shows a further column of values, **Logcap**. It contains the logarithms to base 10 of the **Capital** values. To obtain these you should use the formula @LOG(**Capital**), where the argument **Capital** denotes the cell address of the first **Capital** value. Copying your formula will complete the table, and you can display a Line Graph of **Logcap** for each **Year**, as shown in fig.16.3. Notice that this graph forms a straight line.

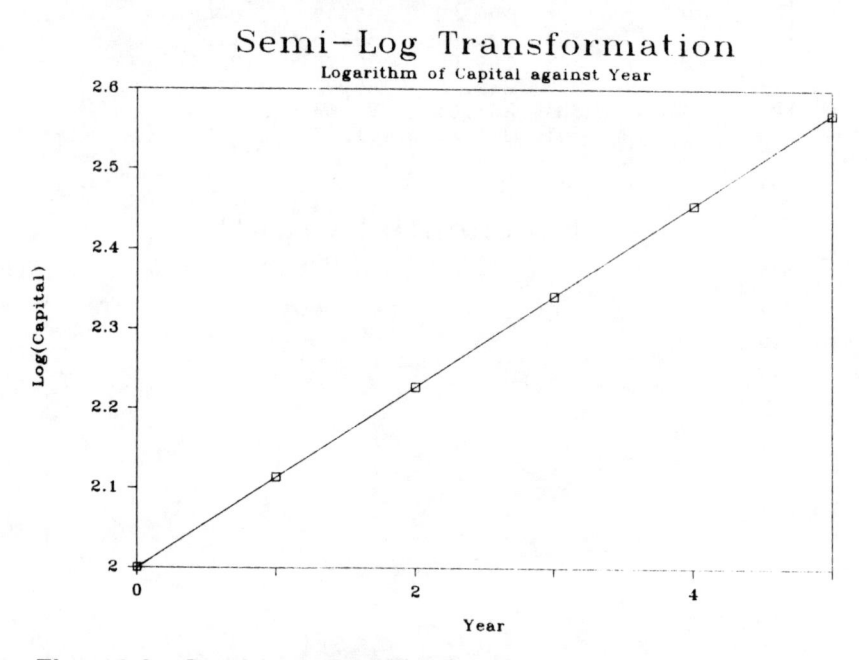

Fig. 16.3 Semi-Log Transformation

16.4 Double Log Transformations

The relationship

$$Y = aX^b \qquad [16.3]$$

is linear in the logarithms of the two variables, **X** and **Y**, since, taking logarithms of both sides we have:

$$\log(Y) = \log(a) + b \log(X) \qquad [16.4]$$

For any particular values of **a** and **b**, graphing **X** against **Y** as an XY graph will give a curve. If **b** is less than 1 it will be concave downwards as in fig. 16.4, and if **b** is greater than 1 it will be concave upwards.

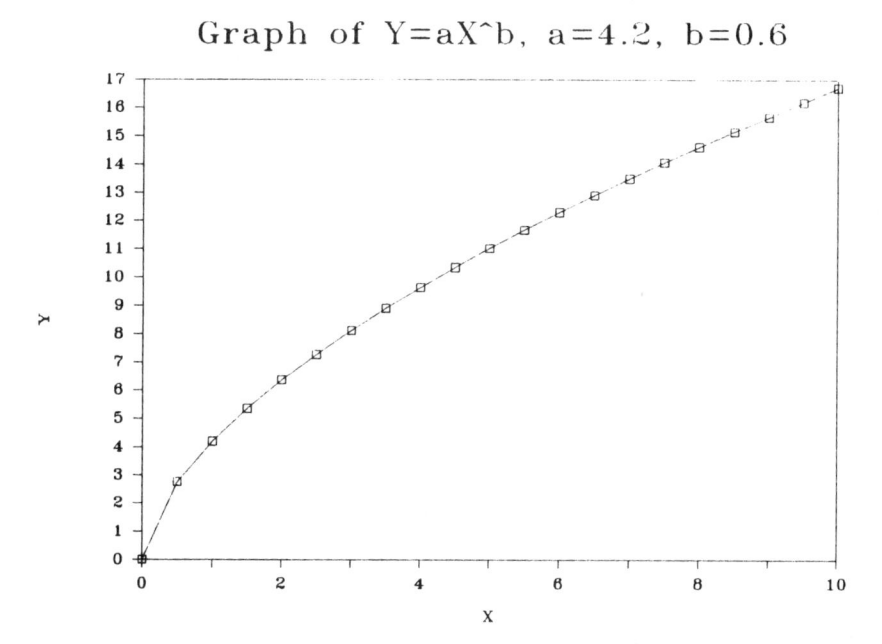

Fig. 16.4 Graph of Y = aX^b, a = 4.2, b = 0.6

A spreadsheet from which fig. 16.4 can be graphed is shown in table 16.3. To construct it, enter and Range Name values for **a** and **b**, and set up a column of **X** values using Data Fill. A formula can now be entered and copied to construct the **Y** values as **a*X^b**, where **X** denotes a relative cell address. Graphing **X** against **Y** will give fig. 16.4. Try altering the values of **a** and **b** and watch the shape of your graph change.

Now construct columns of values of @LOG(**X**) and @LOG(**Y**), again using cell addresses for the variables. An XY Graph of LOG(**X**) against LOG(**Y**) will give a straight line graph of the form shown in fig. 16.5.

Table 16.3 Double Logarithmic Transformation

	A	B	C	D
1	DOUBLE LOGARITHMIC TRANSFORMATION			
2				
3	a	b		
4	4.2	0.6		
5				
6		aX^b		
7	X	Y	@LOG(X)	@LOG(Y)
8	0.0	0.0000		
9	0.5	2.7710	-0.3010	0.4426
10	1.0	4.2000	0.0000	0.6232
11	1.5	5.3568	0.1761	0.7289
12	2.0	6.3660	0.3010	0.8039
13	2.5	7.2780	0.3979	0.8620
14	3.0	8.1194	0.4771	0.9095
15	3.5	8.9062	0.5441	0.9497
16	4.0	9.6491	0.6021	0.9845
17	4.5	10.3556	0.6532	1.0152
18	5.0	11.0314	0.6990	1.0426
19	5.5	11.6806	0.7404	1.0675
20	6.0	12.3067	0.7782	1.0901
21	6.5	12.9121	0.8129	1.1110
22	7.0	13.4992	0.8451	1.1303
23	7.5	14.0697	0.8751	1.1483
24	8.0	14.6252	0.9031	1.1651
25	8.5	15.1670	0.9294	1.1809
26	9.0	15.6962	0.9542	1.1958
27	9.5	16.2137	0.9777	1.2099
28	10.0	16.7205	1.0000	1.2232

16.5 Summary

This chapter gives just a few examples of data transformations. Starting with a column of values of the original variable, it is easy to enter and copy a formula to construct a column of values of the transformed variable. 1-2-3's graphical facilities allow you firstly to look at the original data to assess whether a transformation is likely to be helpful, and secondly to plot the transformed data to see whether the transformation you have chosen has had the desired effect.

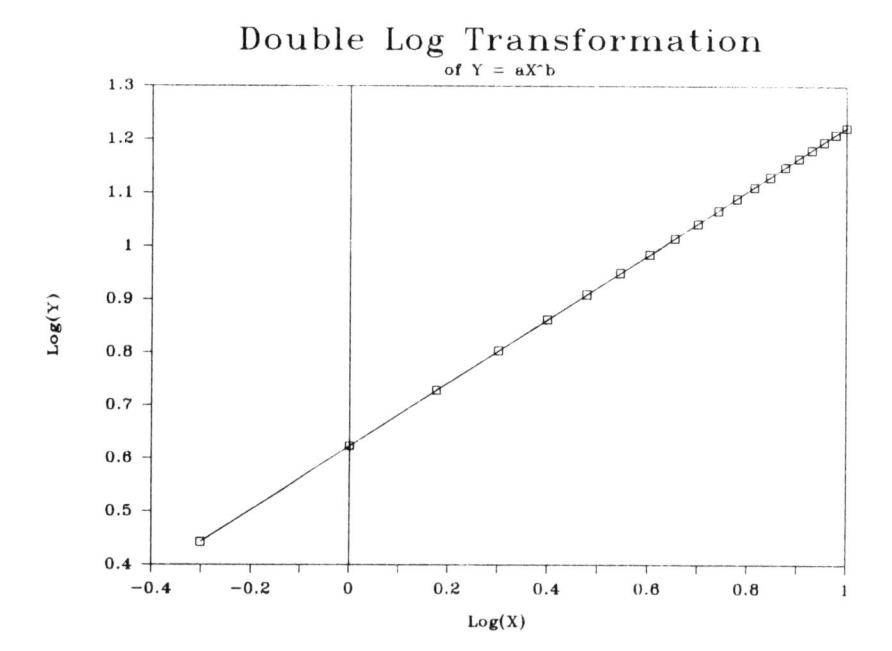

Fig. 16.5 Double Log Transformation

17
Linear Programming

17.1 Introduction

Linear programming is used to determine the optimal use of scarce resources. For instance, we may choose the product mix which, while not using more resources than are available, gives maximum profit. In this chapter we show how to set up a simple problem and find a graphical solution. The Simplex method provides a more general algebraic approach. Arganbright (1985) demonstrates it using a spreadsheet.

17.2 Setting Up the Problem

A typical problem is that of a firm producing both TV and radio sets. Let us assume that from its cost and price data the firm knows that each TV set it sells will yield a profit of £20, and each radio set a profit of £1.50. Normally the firm's output per period is limited by the scarcity of skilled labour and warehouse capacity, but a recent strike has reduced supplies of a certain transistor so that for the current period the number available is limited. For each TV set specific amounts of labour, transistors and warehouse space are required, and for each radio set a different combination of these, as shown in table 17.1.

How many TVs and how many radios should the firm manufacture to maximise its profits?

Table 17.1 Resources Required and Available

Scarce resources	Requirements per set TV	Radio	Available per time period
Labour (man hours)	10	0.5	10,000
Transistors	15	3	22,500
Warehouse space (m^3)	0.5	0.05	2,000

We must define the firm's profits, which are derived from cost and price data. Let **T** be the number of TV sets and **R** the number of radios which the firm produces. The firm's profits, **Z**, then, are

$$Z = 20\,T + 1.5\,R$$

This is called the **objective function**, because we wish to choose **T** and **R** so as to maximise it. Clearly, **Z** becomes larger as **T** and/or **R** becomes larger.

The scarcity of labour, warehouse capacity and transistors, together with the technological requirements of the combinations in which they are to be used, restrict the number of TVs and radios that can be produced per time period. These restrictions can be set out as constraints, one for each scarce factor. Each TV set produced requires 10 man hours of labour and each radio set 0.5 man hours. If, then, **T** TVs and **R** radios are produced, 10T + 0.5R man hours of labour will be used. But the firm cannot use more labour than the 10,000 man hours it has available, so its production is subject to the constraint

$$10\,T + 0.5\,R \leq 10{,}000$$

Similarly, from the transistors required and available we have

$$15\,T + 3\,R \leq 22{,}500$$

and from the warehouse space

$$0.5T + 0.05R \leq 2{,}000$$

A further restriction on the possible values for **T** and **R** is that they may not be negative, since this would imply dismantling sets which had just been produced. We must formally incorporate this in our model, and so have the **non-negativity** conditions

$$T \geq 0,\ R \geq 0.$$

If the optimal solution gives non-integral values for **T** and **R**, the solution values can be rounded to determine the number of items to be produced in a particular time period.

17.3 Graphical Solution

When, as in this problem, there are just two products, a graphical solution can be found as to how much of each should be produced. **T** and **R**, the quantities of TVs and radios manufactured, form the axes. The various combinations of TVs and radios which it is physically possible to produce can be seen by plotting the **constraints**. For each constraint, we plot the line at which it becomes fully effective. Points on, below or to the left of the line, then, represent output combinations which satisfy that constraint. Points which satisfy all of the constraints in this way are **feasible** output combinations.

To find which of these feasible output combinations yields the highest profit, we add **profit lines** to the diagram. Each of these depicts the various combinations of outputs with which a particular profit level can be attained. Since the profit from a TV set is always 20/1.5 of the profit from a radio, the slope of the profit lines is the same regardless of what profit level we are considering, and so a series of parallel lines can be drawn corresponding to different levels of profit.

We wish to choose the output combination that yields the highest profit, and so we look for the highest profit line that just touches the boundary of the feasible output combinations, which occurs in fig. 17.1 at point **M**.

Table 17.2 Data Entries for Linear Programming Problem

	A	B	C	D	E	F
1		LABOUR	TRANS	SPACE	PROFIT	
2	TV	10	15	0.5	20	
3	RADIO	0.5	3	0.05	1.5	
4	AVAILABLE	10000	22500	2000		

To draw the graph using 1-2-3, first enter the data in the spreadsheet. The layout shown in the table 17.2 is convenient for setting up the graphical solution. Let us choose to plot **T**, the number of TVs produced, on the **X** axis. We have to enter a column of values which **T** might take. From the constraints we see that, if no radios are produced, the warehouse constraint will allow 4,000 TVs to be produced, while the other constraints will not permit the production of so many. The highest number we need consider for **T** is, therefore, 4,000; the lowest is, of course, 0. We can use Data Fill to enter the values from 0 to 4000 in steps of 500 in column A. As we find that more detail would be useful at specific points, we shall find that we can insert additional values, as shown in table 17.3.

We shall be plotting six XY graphs, three depicting the constraints and three representing lines of equal profit. By substituting the **T** values in the equations for each we must construct six columns containing **R** values. The equation at which the labour constraint is fully effective is

$$10\,\mathbf{T} + 0.5\,\mathbf{R} = 10{,}000$$

This can be rewritten as

$$\mathbf{R} = 10{,}000\,/\,0.5 - (10\,/\,0.5)\,\mathbf{T}$$

and we enter in column B a formula representing the right hand side into our spreadsheet. The formula should contain references to cells holding constants rather than the constants themselves, so that recalculation is possible if we wish to alter one of the constants displayed. We shall want to copy the formula down the column, so row references to cells containing constants must be in absolute form. The equations for the transistors and warehouse constraints must similarly be rewritten as expressions for **R**, and formulae entered in columns C and D to calculate them.

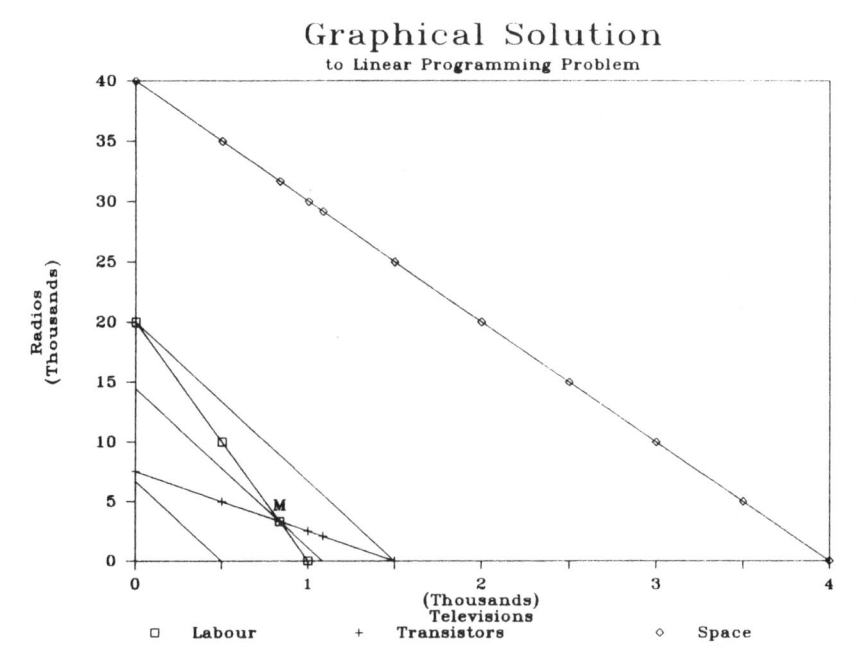

Fig. 17.1   Linear Programming - Graphical Solution

For the equal profit lines we make a guess at what profit is likely to be. 1,000 TV sets would give a profit of £20,000, so let us take £10,000, £20,000 and £30,000 as possible values. We enter these into cells of the spreadsheet so that it will be easy to alter them if we wish. The first profit function is, then,

$$20\,\mathbf{T} + 1.5\,\mathbf{R} = 10,000$$

Again this must be rewritten as an expression for **R**, giving

$$\mathbf{R} = 10,000 / 1.5 - (20 / 1.5)\,\mathbf{T}$$

This formula is entered in column E using absolute row and column references for the profit coefficients and an absolute row address for the profit value. It can then be copied to form two further columns corresponding to the other profit values. When all the formulae have been entered they can then be copied simultaneously down the six columns to form the required sets of **R** values.

We can now plot the graph of these lines, view it, and then make whatever alterations are required. Note that we must define the ranges to be plotted to include only the positive values of **R**. Appropriate titles, legends and formats can be chosen from the graph options.

The line representing £20,000 profit will be found to be just below the intersection of the labour and transistors constraints. Clearly the maximum profit will be given by a line passing through this intersection, and we increase the value from £20,000 until we find the appropriate line. To help us recognise when we have found it we need another value of **T**, rather less than 1,000, which we shall adjust until it takes the value at the intersection of the constraints. We insert a blank row, enter a **T** value, say 900, and copy the formulae into the remaining columns. We then alter the **T** value until the **R** values it gives for the labour and transistors constraints are equal. We now return to the profit value and alter it until it gives the same **R** value as the labour and transistors constraints for that value of **T**. It will be useful to label on the graph the point at which these lines intersect. This can be done by entering a column of data labels, although in this case all but the one corresponding to the point to be labelled are blank. Graph Options Data-Labels is then used.

On plotting the graph now, we find that the profit line we have adjusted does not quite reach the **T** axis. To remedy this we again need to insert another value of **T**, this time a little larger than 1000, and adjust it until the corresponding **R** value for the second profit column is 0. We define the range for which that column is plotted and the graph is complete. The set of data used for it is shown in Table 17.3.

Table 17.3 Calculations for Plotting Graphical Solution

	A	B	C	D	E	F	G
1		LABOUR	TRANS	SPACE	PROFIT		
2	TV	10	15	0.5	20		
3	RADIO	0.5	3	0.05	1.5		
4	TIME	10000	22500	2000			
5					10000	21666.6	30000
6	T	LABOUR	TRANS	SPACE	PROFIT1	PROFIT2	PROFIT3
7	0	20000	7500	40000	6666.7	14444.4	20000.0
8	500	10000	5000	35000	0.0	7777.7	13333.3
9	833	3333.33	3333.33	31666.66	-4444.4	3333.3	8888.9
10	1000	0	2500	30000	-6666.7	1111.1	6666.7
11	1083	-1666.6	2083.35	29166.7	-7777.7	0.0	5555.6
12	1500	-10000	0	25000	-13333.3	-5555.6	0.0
13	2000	-20000	-2500	20000	-20000.0	-12222.3	-6666.7
14	2500	-30000	-5000	15000	-26666.7	-18888.9	-13333.3
15	3000	-40000	-7500	10000	-33333.3	-25555.6	-20000.0
16	3500	-50000	-10000	5000	-40000.0	-32222.3	-26666.7
17	4000	-60000	-12500	0	-46666.7	-38888.9	-33333.3

We can read the values of the optimal solution from table 17.3. The **R** values for the labour and transistors constraints are equal at 3333.3 in row 9. These constraints intersect, therefore, at the point where approximately 3333 radios are produced. Reading the value in the **T** column shows us that the associated production of TVs is 833 sets. We have adjusted the profit figure from which the **R** values in the Profit2 column are calculated so that they represent a line which intersects the labour and transistors constraints. This line represents a profit of £21,666, which is the maximum value. Notice from the graph that the warehouse constraint does not operate: there is excess warehouse space available.

17.4 Summary

With 1-2-3 it is easy to study the graphical solution to a simple linear programming problem. The concepts of feasible output combinations and of the optimal solution are more easily grasped from a diagrammatic representation of the problem. The spreadsheet calculations set up to obtain this allow the solution values for the variables to be read quite accurately.

18
Multiple Regression and Matrices

- Regression: inbuilt facility
- Matrix transposition
- Matrix multiplication
- Matrix inversion
- Conversion of formulae to values

18.1 Introduction

In Chapter 12 we investigated the relationship between two variables by finding the linear regression of **Y** on **X**. **Multiple regression** is an extension of this, where we now specify that the dependent variable, **Y**, is linearly related to the independent variables $X_2, X_3, ... X_k$, so that the regression equation takes the form

$$Y = b_1 + b_2 X_2 + b_3 X_3 + ... + b_k X_k \qquad [18.1]$$

We estimate the **k** regression coefficients, $b_1, b_2, b_3, ... b_k$, from a data set of **n** observations. Each observation must consist of a value for the dependent variable, **Y**, and corresponding values for each of the **k** - 1 independent variables. The goodness of fit of the equation is measured by the square of the multiple correlation coefficient, R^2. This shows the proportion of the variability in the **Y** values explained by fitting the regression equation, as r^2 does for a simple regression. The standard errors of the coefficients are needed to test whether each term in the equation makes a significant contribution to the explanation of the dependent variable, and for determining confidence intervals for the coefficients.

This chapter demonstrates the estimation of a regression on two independent variables, using a Lotus 1-2-3 command first available in Release 2, and shows also how to calculate some useful extra statistics which are not found automatically. One of these is the standard error of the constant term, in the calculation of which we shall utilise **matrix** facilities. We shall also demonstrate the use of the **Range Value** command to retain certain values for comparison purposes in the course of a "What if?" calculation.

18.2 Lotus 1-2-3 Release 2

- **Data Regression**
- **Range Transpose**
- **Data Matrix Multiply**
- **Data Matrix Invert**
- **Range Value**

Release 2 of Lotus 1-2-3 includes a number of commands which are not available in Release 1. The **Data Regression** command estimates a multiple regression equation; **Range Transpose**, **Data Matrix Multiply** and **Data Matrix Invert** provide matrix manipulations; and **Range Value** transforms numerical formulae into values as it copies them.

To use the **Data Regression** command for estimating the multiple regression equation

$$Y = b_1 + b_2 X_2 + b_3 X_3 + ... + b_k X_k \qquad [18.1]$$

the data should be entered using columns for the variables and rows for the observations. Each data row will then comprise a set of values for each of the variables $Y, X_2, X_3, ... X_k$. Up to 16 independent variables can be included; they must be placed in adjacent columns. After selecting **Data Regression** you should select **Y-Range** and define this to be the column of **n** values of the dependent variable. You must also choose **X-Range** to define the independent variables. This range will have $k - 1$ columns and the same number of rows, **n**, as the **Y** range. It is assumed, by default, that you wish to include the constant term, b_1, in your regression, but you can choose not to do so by using the **Intercept Zero** command.

Before issuing the **Go** command to perform the regression, you must specify where the results are to be placed with the **Output-Range** command. As usual when specifying Output or Copy TO ranges in Lotus 1-2-3, it is only necessary to give the top left hand corner of the range. It is important, however, that this cell should be within a sufficiently large blank area of the spreadsheet. The Go command writes output to certain cells, overwriting anything they contain. For readability, nothing is entered in other cells, which are intended to be left blank. They are not, however, erased by the command, so anything you have left in them will contaminate your regression output.

Having defined what you want to do, choose **Go** to activate the regression command. It calculates the coefficients of the equation, b_1, b_2, b_3, ... b_k, and also the standard errors (but not that for the constant term), together with the square of the multiple correlation coefficient, R^2.

This inbuilt facility can, of course, be used for the simple regression equation

$$Y = a + b X \qquad [12.1]$$

The results output by the command are then the coefficients **a** and **b**, together with the standard error of **b**, the square of the simple correlation coefficient, and the standard error of estimate.

Lotus 1-2-3 Release 2 provides facilities for matrix transposition, multiplication and inversion. A **matrix** is simply a rectangular array of cells, such as might be defined as a **range** in Lotus 1-2-3. When using the matrix commands you define ranges to which the matrix operations are to be applied, and output ranges where the results will be placed. Again, the top left hand corner of a blank area of the spreadsheet comprises a suitable output range.

Transposition is performed using the **Range Transpose** command. When a matrix is transposed, it is copied with the rows and columns interchanged. This command should not be applied to cells containing formulae with relative addresses, since the addresses are not adjusted. Notice that if a matrix containing **n** rows and **k** columns is transposed, the resulting matrix will have **k** rows and **n** columns.

Two matrices can only be **multiplied** if the number of columns in the first equals the number of rows in the second. The resulting matrix then has the number of rows of the first matrix, but the number of columns of the second. Lotus 1-2-3 incorporates a size restriction on matrices to be multiplied, namely that the number of rows in the first matrix multiplied by the number of columns in the second should be less than 8192.

Appropriate matrices can be multiplied using the **Data Matrix Multiply** command.

Multiplication by the **inverse** of a matrix is the equivalent, in matrix algebra, to division. The product of a matrix and its inverse gives the identity matrix. This has ones in the principal diagonal, and zeros elsewhere. Not all matrices have an inverse. Those that do are square, and their determinant (defined in any book covering matrix algebra, e.g. Curwin & Slater (1985)) is not equal to zero. The inverse of invertible matrices can be found in Lotus 1-2-3 using the **Data Matrix Invert** command.

A further feature of Lotus 1-2-3 Release 2 is the ability to **copy calculated values**, rather than the formulae which computed them. This is done using the **Range Value** command. The values can be copied to a different part of the spreadsheet, leaving the original formulae intact, or they can replace the latter.

18.3 Multiple Regression

- ### Data Regression

As a simple example of **multiple regression**, we shall investigate whether the quantity of maize harvested in the U.K. (measured in thousands of tonnes) depends on the area of maize planted (measured in hectares) and on the number of millimetres of rain falling in June, July and August. Since maize requires a substantial amount of water for growth, we might expect a positive relationship with both these variables. Columns B to D of table 18.1 contain the values of these three variables for the years 1973 to 1986, which constitute the observations. Notice that the two independent variables, Maize Area and Rain, have been entered in the adjacent columns C and D.

After the data have been entered, move the cell pointer to the top left hand corner of a blank area of the spreadsheet, ready to define that as the **regression output range.** The cell E24 has been chosen in the spreadsheet shown. A cell in columns A to D would not be suitable in this instance, because those columns have been narrowed in width, and there would not be space for the labels output by the command to be fully displayed.

Table 18.1 Multiple Regression Data

```
        A         B         C         D
1  MAIZE HARVESTED ('000 tonnes) with
2  AREA PLANTED ('000 hectares) and
3  RAINFALL IN JUNE, JULY & AUGUST (mm)
4  [Source: Annual Abstract of Statistics]
5            Harvest    Area      Rain
6  ----------------------------------
7  1973       336         7        218
8  1974       609        16        238
9  1975       887        26        139
10 1976       839        29         76
11 1977      1198        35        211
12 1978      1017        26        230
13 1979       895        25        169
14 1980       785        22        298
15 1981       635        18        152
16 1982       635        16        258
17 1983       550        15        110
18 1984       580        16        127
19 1985       770        20        284
20 1986       915        23        214
```

You are now ready to define the regression to be performed. Obtain the menu and choose **Data** followed by **Regression**. Select **Y-Range** and define this as the cells containing values in the Maize Harvest column, namely B7..B20. After entering this, choose **X-Range** and point to the values in the Maize Area and Rain columns, marking out and then entering the range C7..D20. Enter as the **Output-Range** the cell E24 to which you moved the pointer, and select **Go** to activate the regression command. The regression output produced will be that shown in cells E24 to H32 of table 18.2. Identifying labels have been added in column I, namely = n for the number of observations, and = n - k for the number of degrees of freedom.

The results obtained enable us to write down the regression equation in the usual format:

$$Y = -6.764 + 30.679\, X_2 + 0.634\, X_3 \qquad [18.2]$$
$$ (2.056)\quad (0.217)$$
$$R^2 = 0.953$$

From this, the remaining figures shown in table 18.2 can be computed.

To test whether each of the independent variables contributes significantly to the explanation of the **Y** values, **t** values, shown as 14.923 and 2.925, are computed in cells G33 and H33. These values are computed by dividing the **X** coefficients two rows above by their respective standard errors. The **t** values have the number of degrees of freedom shown in cell H29, namely 11, and **t** tests can be carried out as described for a simple regression in section 12.6. Referring to **t** tables we find $t_{0.01} = 2.72$, hence both coefficients are significantly different from zero at the 0.01 significance level, and we conclude that both the maize area planted and the summer rain influence the quantity of maize harvested.

Table 18.2 Multiple Regression Output and Residuals

```
           E              F              G              H              I
 4    Estimated
 5    Harvest            e            ediff           e^2          ediff^2
 6    -----------------------------------------------------------------------
 7     346.12         -10.12                         102.38
 8     634.91         -25.91         -15.79          671.08          249.23
 9     878.97           8.03          33.93           64.44         1151.44
10     931.09         -92.09        -100.12         8481.20        10024.22
11    1200.71          -2.71          89.39            7.33         7989.92
12     936.63          80.37          83.08         6459.22         6901.68
13     867.30          27.70         -52.67          767.22         2774.18
14     857.00         -72.00         -99.70         5183.75         9939.51
15     641.77          -6.77          65.22           45.88         4254.23
16     647.58         -12.58          -5.80          158.19           33.68
17     523.12          26.88          39.45          722.32         1556.57
18     564.57          15.43         -11.45          237.94          131.12
19     786.77         -16.77         -32.19          281.20         1036.46
20     834.45          80.55          97.31         6487.55         9470.06
21    -----------------------------------------------------------------------
22                      0.00                       29669.71        55512.31
23
24                        Regression Output:
25    Constant                                       -6.764
26    Std Err of Y Est                               51.935
27    R Squared                                       0.953
28    No. of Observations                              14 = n
29    Degrees of Freedom                               11 = n-k
30
31    X Coefficient(s)         30.679                 0.634
32    Std Err of Coef.          2.056                 0.217
33    t values                 14.923                 2.925
34
35    No. of Indep. Variables                           2 = k-1
36    F                                              111.945
37    Adjusted R Squared                               0.945
38    Durbin-Watson d-statistic                        1.871
```

The number of independent variables, **k** - 1, has been entered in cell H35, and is shown as 2. Notice that the number of degrees of freedom is **n** - **k**, so if repeated use of the Regression command is planned it is convenient to compute **k** - 1 by a formula using two automatically calculated results, namely as **n** - (**n** - **k**) - 1.

The **F** statistic calculated in cell H36 and displayed as 111.945 is, like the **F** ratios calculated in Chapter 13, a ratio of two mean sums of squares. The one in the numerator shows the variance explained by fitting the regression model, and that in the denominator shows the random variance that is left unexplained by the model which has been fitted (see Thomas (1983)). It can be defined as:

$$F = [R^2 / (k - 1)] / [(1 - R^2) / (n - k)] \qquad [18.3]$$

and a formula to compute the right hand side of this expression is entered in cell H36. If the regression model has explained little of the variation in the **Y** values, **F** will be close to 1, while if the model is a good fit the **F** value will be large. Our **F** value has **k - 1** and **n - k** degrees of freedom, and we can use it to conduct a joint test on all the independent variables in the regression, which is equivalent to saying we can test whether the multiple correlation coefficient is zero. Tables show $F_{0.01}$ to be 7.206, and since our **F** value is larger than this we can reject the hypothesis that the independent variables, considered jointly, have no influence on the variable, and equivalently reject the hypothesis that the multiple correlation coefficient is zero.

If another regression model were to be fitted to the data including additional independent variables, the new value of R^2 could not be less than the one already obtained. Each additional independent variable, however, reduces by 1 the number of degrees of freedom available. To take this into account an adjusted value of R^2 can be computed, as is done in cell H37. The adjusted value is defined as follows:

$$\text{Adjusted } R^2 = 1 - (1 - R^2)(n - 1) / (n - k) \qquad [18.4]$$

It is possible for this adjusted R^2 value to decrease as further variables are added to a regression model if the increase in R^2 is not sufficient to offset the decrease in the value of **n - k**. By entering a formula for the right hand side of the above expression in cell H37, the value of 0.945 is obtained.

The Estimated Harvest values shown in column E are found by successively substituting each pair of X_2 and X_3 values in the regression equation

$$Y = -6.764 + 30.679\, X_2 + 0.634\, X_3 \qquad [18.2]$$

To do this, then, we enter an appropriate formula at the top of the Estimated Harvest column, using absolute addresses for the coefficients and relative ones for the Maize Area and Rain values. The formula is then copied down the column, giving the values shown in table 18.2.

The regression model makes certain assumptions about the residuals, and it may be useful to examine the pattern of residuals obtained to check whether these assumptions appear to hold. Accordingly, the residuals, **e**, are calculated in column F of table 18.2, by entering and copying a formula to find the difference between the Maize Harvest values in column B and the Estimated Harvest values in Column E.

The Durbin-Watson **d**-statistic is used when dealing with time series data to test whether successive residuals are related to one another, thus violating the assumption that they are random. The calculation of this statistic involves the computation of the differences between successive residuals, $e_t - e_{t-1}$. The formula entered in row 8 of column G therefore subtracts the residual in the previous row, e_{t-1}, from that in the current row, e_t, giving, when it is copied down, the column of values titled ediff. Squares of the residuals and of the differenced values are computed in columns H and I respectively. The column sums formed in H22 and I22 are then $\sum e_t^2$ and $\sum (e_t - e_{t-1})^2$ respectively. The Durbin-Watson **d**-statistic is defined as

$$\mathbf{d} = \sum (e_t - e_{t-1})^2 / \sum e_t^2 \qquad [18.5]$$

and hence it is computed in cell H38 as the ratio of the value in I22 to that in H22, giving the value of 1.871. Random residuals yield a Durbin-Watson statistic of about 2. Tables show for various combinations of numbers of independent variables and of observations, the d values below which, at the specified significance level, there is evidence of non-randomness. For 2 independent variables and 14 observations the tabulated value is 1.55 at the 0.05 significance level. Our value is greater than the tabulated value, and therefore we may reasonably assume that successive values of our residuals are not related to one another.

18.4 The Variance-Covariance Matrix

- **Range Transpose**
- **Data Matrix Multiply**
- **Data Matrix Invert**

The terms variance and covariance were defined in section 12.3. In multiple regression analysis when k coefficients $b_1, b_2, b_3, ... b_k$ have been estimated, the variance-covariance matrix is a square array of **k** rows and **k** columns. In the principal diagonal (from top left to bottom right) it contains the variances of the coefficients. That is, the variance of b_3, which is the square of the standard error of **b**$_3$, is found in row 3 and column 3 of the matrix. The off-diagonal elements give the covariances of the various pairs of coefficients, each of which appears twice. For example, the value in row 2 and column 3 of the matrix, and also that in row 3 and column 2, gives the covariance of **b**$_2$ and **b**$_3$. We shall find the variance-covariance matrix for the regression which we have performed.

We begin by defining the **X** matrix. This comprises the columns of values of the independent variables, preceded by a column of units, to allow for the constant term in the regression. Our **X** matrix, then, has **n** rows and **k** columns, where in this instance **n** = 14 and **k** = 3. Table 18.3 displays it in cells N7 to P20. The column of 1's is obtained using Data Fill, and columns O and P are copied from columns C and D.

Table 18.3 Matrix X and its Transpose X'

	N	O	P	Q	R	S	T	U	V	W	X	Y	Z	AA
1	X' Matrix
2	1	1	1	1	1	1	1	1	1	1	1	1	1	1
3	7	16	26	29	35	26	25	22	18	16	15	16	20	23
4	218	238	139	76	211	230	169	298	152	258	110	127	284	214
5	
6	X Matrix
7	1	7	218
8	1	16	238
9	1	26	139
10	1	29	76
11	1	35	211
12	1	26	230
13	1	25	169
14	1	22	298
15	1	18	152
16	1	16	258
17	1	15	110
18	1	16	127
19	1	20	284
20	1	23	214

We require also **X'**, the transpose of the matrix we have just obtained. **X'** has the rows and columns of matrix **X** interchanged, so it will occupy **k** rows and **n** columns. Transposition is performed in Release 2 of Lotus 1-2-3 using the **Range Transpose** command. This asks for a range to Copy FROM, when you should move the pointer to cover the **X** matrix just set up in N7..P20. After entering this range, you will be asked for a range to Copy TO. Enter a cell at the top left hand corner of a sufficiently large blank area, such as N2. On pressing [RETURN] the transposition is performed, and the values appear as shown in cells N2 to AA4 of table 18.3.

The next step is to form the matrix product **X'X**. We note that there are 14 columns in matrix **X'**, the same as the number of rows in matrix **X**. It is therefore possible to form the product **X'X**, and that matrix will have 3 rows and 3 columns. Note that it would also be possible to form the product **XX'**, but this would be a different matrix with 14 rows and columns. Matrix multiplication is performed by the **Data Matrix Multiply** command. On being asked for the first range to multiply, you should enter the range containing **X'**, namely N2..AA4. In respose to the request for a second range to multiply, you should enter the **X** range, N7..P20. You will again be asked to indicate an output range. If you reply K26, the product will be formed as shown in table 18.4. For our **X** matrix, the **X'X** matrix is as follows:

$$\mathbf{X'X} = \begin{matrix} n & \Sigma x_2 & \Sigma x_3 \\ \Sigma x_2 & \Sigma x_2^2 & \Sigma x_2 x_3 \\ \Sigma x_3 & \Sigma x_2 x_3 & \Sigma x_3^2 \end{matrix} \quad [18.5]$$

The elements of this matrix can be identified, then, as the sums of the values of the independent variables, and of their squares and cross products.

We now require the inverse of this matrix, denoted $(\mathbf{X'X})^{-1}$. We shall find this using the Lotus 1-2-3 Release 2 command **Data Matrix Invert**. We define as the Range to invert the cells containg the **X'X** matrix, namely K26..M28, and give a suitable cell as the Output range. The address K31 has been given for this to produce the output shown in table 18.4. You can, of course, check that the product of **X'X** and $(\mathbf{X'X})^{-1}$ is the identity matrix, by multiplying them using the Data Matrix Multiply command.

We can now find the variance-covariance matrix, which is defined as

$$\text{var}(\mathbf{b}) = \sigma^2 (\mathbf{X'X})^{-1} \quad [18.6]$$

where σ^2 is estimated by the standard error of the estimate of **Y**, which is one of the results produced by the Data Regression command. The variance-covariance matrix shown in table 18.4, then, is found by multiplying each element in the $(\mathbf{X'X})^{-1}$ matrix by the square of the value in cell H26, as shown in table 18.2. This yields the matrix shown in cells K36 to M38. Taking the square root of the diagonal elements gives the standard errors of the regression coefficients, as shown in cells K41 to M41. The first of these, displayed as 65.428, is the standard error of the constant term, b_1. The other two are the standard errors of the **X** coefficients, b_2 and b_3. These values of 2.056 and 0.217 had already been found automatically by the Data Regression command.

Table 18.4 Product, Inverse and Var-Covar Matrix

```
            K           L          M
25  X'X  Matrix
26          14         294       2724
27         294        6822      56446
28        2724       56446     588400
29
30  Inverse [X'X]
31    1.58712   -0.03687   -0.00381
32   -0.03687    0.00157    0.00002
33   -0.00381    0.00002    0.00002
34
35  Variance-Covariance Matrix
36   4280.841     -99.435    -10.279
37    -99.435       4.227      0.055
38    -10.279       0.055      0.047
39
40  Standard Errors
41      65.428       2.056      0.217
```

18.5 Retaining Values in Calculations

- **Range Value**

In section 7.4 we described how a formula could be replaced by its numerical value, using the facility available in Lotus 1-2-3 Release 1. An easier way of doing this is to use the **Range Value** command provided in Release 2. This is used like the Copy command, so anything you wish to copy, including labels, can be included in the Copy FROM range. With Range Value, however, as the copying is done all formulae are replaced by their numerical values. Full accuracy is retained for these values, not just the number of decimal places displayed. The Copy TO range is defined in the usual way, except that with Range Value it is possible for this to be the same as the Copy FROM range. This means you can either leave your original formulae intact and ready to use in a "What if?" calculation while retaining a copy of a set of values they produced, or you can overwrite your formulae with the corresponding numerical values.

As an example of the use of this command, we shall retain, for comparison purposes, the residuals that we have obtained in our multiple regression, while performing a simple regression of Maize Harvest on Maize Area to investigate the effect of ignoring the Rain data. Remember to save your completed spreadsheet before starting to modify it. We use the Range Value command, then, to copy the residuals from column F into column J, where they are given the new title, Mult Regn Residuals, as shown in table

18.5. The results of the previous Regression command are then Range Erased, but not the further statistics calculated from them. Blank cells have the numerical value zero, so the column of estimated values all become zero. The residuals and all values calculated from them are meaningless, and some of the statistics we calculated display ERR. Data Regression is used again, but this time the X-Range is defined only to consist of the Maize Area values. Choosing an Output-Range of E24 as before and selecting Go, the results are produced as shown in table 18.5. Notice that the correct further statistics are all displayed as well. As only one **X** coefficient has been estimated, the **t** value corresponding to the second, which displayed ERR, has been Range Erased.

Table 18.5 Simple Regression using Data Regression

	E	F	G	H	I	J
4	Estimated					Mult Regn
5	Harvest	e	ediff	e^2	$ediff^2$	Residuals
6	---	---	---	---	---	---
7	341.65	-5.65		31.92		-10.12
8	611.09	-2.09	3.56	4.39	12.64	-25.91
9	910.48	-23.48	-21.38	551.17	457.22	8.03
10	1000.29	-161.29	-137.81	26015.07	18992.92	-92.09
11	1179.92	18.08	179.37	326.83	32173.73	-2.71
12	910.48	106.52	88.44	11347.13	7822.42	80.37
13	880.54	14.46	-92.06	209.13	8475.36	27.70
14	790.72	-5.72	-20.19	32.76	407.44	-72.00
15	670.97	-35.97	-30.25	1293.91	914.88	-6.77
16	611.09	23.91	59.88	571.48	3585.20	-12.58
17	581.16	-31.16	-55.06	970.70	3031.79	26.88
18	611.09	-31.09	0.06	966.86	0.00	15.43
19	730.85	39.15	70.25	1532.92	4934.63	-16.77
20	820.66	94.34	55.19	8899.61	3045.40	80.55
21	---	---	---	---	---	---
22		0.00		52753.89	83853.65	0.00
23						
24	Regression Output:					
25	Constant			132.082		
26	Std Err of Y Est			66.304		
27	R Squared			0.917		
28	No. of Observations			14	= n	
29	Degrees of Freedom			12	= n-k	
30						
31	X Coefficient(s)		29.938			
32	Std Err of Coef.		2.605			
33	t values		11.494			
34						
35	No. of Indep. Variables			1	= k-1	
36	F			132.116		
37	Adjusted R Squared			0.910		
38	Durbin-Watson d-statistic			1.590		

18.6 Summary

Multiple regression can easily be performed using a Lotus 1-2-3 Release 2 command. The results are enhanced by the calculation of further statistics, for some of which matrix computations are needed. Facilities for these are also provided in Release 2 of Lotus 1-2-3. Another Release 2 command makes it easy to retain selected numerical values when performing "What if?" calculations.

19
Past, Present and Future

19.1 Introduction

This book has concentrated on Lotus 1-2-3, but this was neither the first nor is it the latest spreadsheet. In fact most of the applications covered in this book can also be performed in other spreadsheet programs. This chapter begins with the very first spreadsheet, VisiCalc, then covers two second-generation spreadsheets: SuperCalc and MultiPlan. It then looks at 1-2-3 and its successors: integrated software packages. However, the development of the spreadsheet reached its zenith with Lotus 1-2-3 and it has not yet been bettered. The future lies rather in the novel applications to which spreadsheets are increasingly being put. We consider the initial application that prompted the spreadsheet, finance, then its natural extension to mathematics, science and now social science.

19.2 VisiCalc

The very first spreadsheet of all was VisiCalc (VC) invented by Daniel Bricklin and Robert Frankston in 1979. It proved a marketing success and led to enhanced sales of the Apple II microcomputer on which it initially ran. Table 19.1 shows the command equivalents between VisiCalc and 1-2-3.

Table 19.1 VisiCalc Command Equivalents

LOTUS 1-2-3 Command:	Meaning:	VISICALC Command:	Meaning:
/C	Copy	/R	Replicate
/F	File	/S	Storage
/FR	File Retrieve	/SL	Storage Load
/FS	File Save	/SS	Storage Save
/M	Move	/MC	Move Column
/M	Move	/MR	Move Row
/P	Print	/P	Print
/Q	Quit	/SQ	Storage Quit
/RE	Range Erase	/B	Blank
/RF	Range Format	/F	Format
/RFF0	Range Format Fixed 0	/FI	Format Integer
/WD	Worksheet Delete	/D	Delete
/WE	Worksheet Erase	/C	Clear
/WG	Worksheet Global	/G	Global
/WGC	W Global Column-width	/GC	G Column-width
/WGF	W Global Format	/GF	G Format
/WGRA	W Global Recalc Automatic	/GRA	G Recalc Auto
/WGRC	W Global Recalc Column	/GOC	G Order Column
/WGRM	W Global Recalc Manual	/GRM	G RecalcManual
/WGRR	W Global Recalc Row	/GOR	G Order Row
/WI	Worksheet Insert	/I	Insert
/WIC	Worksheet Insert Column	/IC	Insert Column
/WIR	Worksheet Insert Row	/IR	Insert Row
/WT	Worksheet Titles	/T	Title
/WW	Worksheet Window	/W	Window

Table 19.2 Other VisiCalc Equivalents

LOTUS 1-2-3 Option:	Meaning:	VISICALC Option:	Meaning:
@AVG()	Average Function	@AVERAGE()	Average
@HLOOKUP()	Horizontal Look Up	@LOOKUP()	Look Up
@ISERR()	Is an error	@ISERROR()	Is an error
@LOG()	Logs to base 10	@LOG10()	Logs(base 10)
@VLOOKUP()	Vertical Look Up	@LOOKUP()	Look Up
["]	Right aligned label	/FR	Format Right
[']	Left aligned label	/FL	Format Left
[\]	Repeat label prefix	/-	Repeat
[+]	Start Formula	[+]	Start Formula
[1/HELP]	Help key	[?]	Help
[2/EDIT]	Edit key	/E	Edit
[5/GOTO]	Goto key	[>]	Goto
[6/WINDOW]	Window key	[;]	Next Window
[9/CALC]	Calculate key	[!]	Recalculate
#AND#	And	@AND()	And
#NOT#	Not	@NOT()	Not
#OR#	Or	@OR()	Or

Although VisiCalc has column letters and row digits just like 1-2-3, it does not use dollar signs for absolute addresses. Instead a formula is entered and then the computer prompts, for each address in the formula, whether it is [R] a Relative address or [N] a No-change (or absolute) address. As table 19.2 shows, VisiCalc has no concept of label prefixes or function keys.

19.3 SuperCalc

Amongst the second generation of spreadsheets which quickly followed VisiCalc, one of the most popular proved to be SuperCalc (SC) from Sorcim (Micros backwards!), launched in 1980. Table 19.3 shows the command equivalents between SuperCalc and 1-2-3; notice that some facilities are only available in later versions.

Table 19.3 SuperCalc Command Equivalents

LOTUS 1-2-3 Command:	Meaning:	SUPERCALC Command:	Meaning:
/C	Copy	/R	Replicate 1:many
/C	Copy	/C	Copy many:many
/D	Data	//D (v3)	Database
/DS	Data Sort	/A (v2&3)	Arrange
/FC	File Combine	/LP	Load Part
/FR	File Retrieve	/L	Load
/FR	File Retrieve	/LA	Load All
/FS	File Save	/S	Save
/FS	File Save	/SA	Save All
/FX	File eXtract	/SP	Save Part
/GV	Graph View	/V (v.3)	View
/M	Move	/M	Move
/M	Move	/MC	Move Column
/M	Move	/MR	Move Row
/P	Print	/O	Output
/Q	Quit	/Q	Quit
/RE	Range Erase	/B	Blank
/RF	Range Format	/FR	Format Row
/RF	Range Format	/F	Format
/RF	Range Format	/FE	Format Entry
/RF	Range Format	/FC	Format Column
/RFC2	Range Format Currency	/F$	Format Dollar
/RFF0	Range Format Fixed 0	/FI	Format Integer
/RP	Range Protect	/P	Protect
/RU	Range Unprotect	/U	Unprotect
/WCS1-255	Worksheet Column Set	/F0-127	Format
/WD	Worksheet Delete	/D	Delete
/WDC	Worksheet Delete Col.	/DC	Delete Column
/WDR	Worksheet Delete Row	/DR	Delete Row
/WE	Worksheet Erase	/Z	Zap
/WG	Worksheet Global	/G	Global

Table 19.3 SuperCalc Command Equivalents Continued

LOTUS 1-2-3 Command:	Meaning:	SUPERCALC Command:	Meaning:
/WGF	W Global Format	/FG	Format Global
/WGF	W Global Format	/GF	Global Format
/WGRA	W Global Recalc Auto	/GA	Global Automatic
/WGRC	W Global Recalc Col.	/GC	Global Column
/WGRM	W Global Recalc Man.	/GM	Global Manual
/WGRR	W Global Recalc Row	/GR	Global Row
/WI	Worksheet Insert	/I	Insert
/WIC	Worksheet Insert Col.	/IC	Insert Column
/WIR	Worksheet Insert Row	/IR	Insert Row
/WT	Worksheet Titles	/T	Title
/WTB	W Titles Both	/TB	Title Both
/WTC	W Titles Clear	/TC	Title Clear
/WTH	W Titles Horizontal	/TH	Title Horizontal
/WTV	W Titles Vertical	/TV	Title Vertical
/WW	Worksheet Window	/W	Window
/WWC	W Window Clear	/WC	Window Clear
/WWH	W Window Horizontal	/WH	WindowHorizontal
/WWS	W Window Sync.	/WS	Window Synch.
/WWU	W Window Unsync.	/WU	Window Unsynch.
/WWV	W Window Vertical	/WV	Window Vertical

Table 19.4 Other SuperCalc Equivalents

LOTUS 1-2-3 Option:	Meaning:	SUPERCALC Option:	Meaning:
@AVG()	Average Function	AVERAGE()	Average
@ERR	Error	ERROR	Error
@HLOOKUP()	Horizontal Look Up	LOOKUP()	Look Up
@ISERR()	Is an error	ISERROR()	Is an error
@RAND	Random number	RANDOM	Random number
@VLOOKUP()	Vertical Look Up	LOOKUP()	Look Up
["]	Right aligned label	/FTR	Format TextRight
[']	Left aligned label	/FTL	Format TextLeft
[']	Left aligned label	["] (v1&2)	Label prefix
[\]	Repeat label prefix	[']	Repeat
[1/HELP]	Help key	[?]	Help
[2/EDIT]	Edit key	/E	Edit
[5/GOTO]	Goto key	[=]	Goto
[6/WINDOW]	Window key	[;]	Next Window
[9/CALC]	Calculate key	[!]	Recalculate
[ALT]letter	Execute macro	/X	eXecute
{DOWN}	Down arrow	[v]	Down arrow
{LEFT}	Left arrow	[<]	Left arrow
{RIGHT}	Right arrow	[>]	Right arrow
{UP}	Up arrow	[^]	Up arrow
#AND#	And	AND()	And
#OR#	Or	OR()	Or

Although SuperCalc has column letters and row digits just like 1-2-3, it does not use dollar signs for absolute addresses. Instead a formula is entered followed by a comma [,] and then the computer prompts whether you require [N] for No adjustment, [A] meaning Ask for adjust or [V] for Values. If you specify [A] the computer prompts "Adjust?" for each address in the formula to which you can respond either [Y] for Yes or [N] for No.

In table 19.4 notice that the SuperCalc's functions do not begin with [@] and that some facilities are only available in later versions.

19.4 MultiPlan

Also amongst the second generation of spreadsheets which quickly followed VisiCalc was MultiPlan (MP) from Microsoft (MS), launched in 1981. Table 19.5 shows the command differences between MultiPlan and 1-2-3. Notice that the former's commands begin with the [ESC] key rather than [/]

Table 19.5 MultiPlan Command Equivalents

```
LOTUS    1-2-3                    MULTIPLAN
Command: Meaning:                 Command:   Meaning:
/C       Copy                     [ESC]CD    Copy Down one:n
/C       Copy                     [ESC]CF    Copy From n:n
/C       Copy                     [ESC]CR    Copy Right one:n
/C       Copy                     [ESC]C     Copy
/DS      Data Sort                [ESC]S     Sort
/F       File                     [ESC]T     Transfer
/FC      File Combine             [ESC]X     eXternal
/FD      File Directory           [ESC]TL    Transfer Load
/FD      File Directory           [ESC]TO    Transfer Options
/FE      File Erase               [ESC]TD    Transfer Delete
/FR      File Retrieve            [ESC]TL    Transfer Load
/FS      File Save                [ESC]TS    Transfer Save
/M       Move                     [ESC]MR    Move Row
/M       Move                     [ESC]MC    Move Column
/M       Move                     [ESC]M     Move
/P       Print                    [ESC]P     Print
/PF      Print File               [ESC]PF    Print File
/PP      Print Printer            [ESC]PP    Print Printer
/PPOM    Print P Options Margin   [ESC]PM    Print Margins
/PPOOC   Print P O Other Cell-Form.[ESC]PO   Print Options
/PPOS    Print P Options Setup    [ESC]PO    Print Options
/PPR     Print Printer Range      [ESC]PO    Print Options
/Q       Quit                     [ESC]Q     Quit
/RE      Range Erase              [ESC]B     Blank
/RF      Range Format             [ESC]FC    Format Cells
/RF      Range Format             [ESC]F     Format
/RF,     Range Format Comma       [ESC]FO    Format Options
```

Table 19.5 MultiPlan Command Equivalents Continued

LOTUS 1-2-3		MULTIPLAN	
Command:	Meaning:	Command:	Meaning:
/RFC2	Range Format Currency 2	[ESC]F$	Format Dollar
/RFF0	Range Format Fixed 0	[ESC]FI	Format Integer
/RFT	Range Format Text	[ESC]FO	Format Options
/RNC	Range Name Create	[ESC]N	Name
/RP	Range Protect	[ESC]L	Lock
/WCS	Worksheet Column-width Set	[ESC]FW	Format Width
/WD	Worksheet Delete	[ESC]D	Delete
/WDC	Worksheet Delete Column	[ESC]DC	Delete Column
/WDR	Worksheet Delete Row	[ESC]DR	Delete Row
/WE	Worksheet Erase	[ESC]TC	Transfer Clear
/WGCS	Worksheet G Column-wid.Set	[ESC]FDW	Format Def. Width
/WGF	Worksheet Global Format	[ESC]FD	Format Default
/WGR	Worksheet Global Recalc.	[ESC]O	Options
/WI	Worksheet Insert	[ESC]I	Insert
/WIC	Worksheet Insert Column	[ESC]IC	Insert Column
/WIR	Worksheet Insert Row	[ESC]IR	Insert Row
/WT	Worksheet Titles	[ESC]WST	Window Split Title
/WW	Worksheet Window	[ESC]W	Window
/WWC	Worksheet Window Clear	[ESC]WC	Window Close
/WWH	Worksheet Window Horizon.	[ESC]WSH	Window Split Horiz
/WWS	Worksheet Window Sync	[ESC]WL	Window Link
/WWV	Worksheet Window Vertical	[ESC]WSV	Window Split Vert.

Whereas most spreadsheets, including 1-2-3, VisiCalc and SuperCalc, use letters to denote columns, MultiPlan uses RnCm where n and m are row and column numbers respectively, placed in square brackets when in relative mode. So the absolute address $B$3 appears as R3C2, and A1's address from a formula in B3 appears as R[-2]C[-1]. The important things to remember are to use the arrow keys to point to relative addresses, use [:] to anchor a range, and just to type in absolute addresses directly. Table 19.6 shows other MultiPlan equivalents, notice that its functions are not preceded by [@].

Table 19.6 Other MultiPlan Equivalents

LOTUS 1-2-3		MULTIPLAN	
Command:	Meaning:	Command:	Meaning:
@AVG()	Average Function	AVERAGE()	Average
@CHOOSE()	Choose	INDEX()	Index
@FALSE	False	FALSE()	False
@HLOOKUP()	Horizontal Look Up	LOOKUP()	Look Up
@ISERR()	Is an error	ISERROR()	Is an error
@LOG()	Logs to base 10	LOG10()	Logs to base 10
@NA	Not Available	NA()	Not Available
@PI	3.141592	PI()	3.141592
@STD()	Standard Deviation	STDEV()	Stan. Deviation
@TRUE	True	TRUE()	True
@VLOOKUP()	Vertical Look Up	LOOKUP()	Look Up

Table 19.6 Other MultiPlan Equivalents Continued

LOTUS 1-2-3 Command:	Meaning:	MULTIPLAN Command:	Meaning:
["]	Right aligned label	[ESC]FR	Format Right
[']	Left aligned label	[ESC]FL	Format Left
[\]	Repeat label	REPT()	Repeat
[+]	Formula key	[=]	Formula key
[.]	Range anchor key	[:]	Range anchorkey
[1/HELP]	Help key	[ESC]H	Help
[1/HELP]	Help key	[?]	Help key
[2/EDIT]	Edit key	[ESC]E	Edit
[3/NAME]	Name key	[@]	Name key
[3/NAME]	Name key	[ESC]N	Name
[4/ABS]	Absolute key	[@]	Absolute key
[5/GOTO]	Goto key	[ESC]G	Goto
[5/GOTO]	Goto key	[ESC]GR	Goto Row-col
[6/WINDOW]	Window key	[ESC]GW	Goto Window
[6/WINDOW]	Window key	[1/WINDOW]	Window key
[9/CALC]	Calculate key	[4/RECALC]	Recalculate key
[9/CALC]	Calculate key	[!]	Recalculate key

19.5 Integrated Software

Lotus 1-2-3, introduced in 1983, was designed to run on the IBM Personal Computer, whose sales it enhanced just as VisiCalc had helped to sell many Apple II microcomputers. 1-2-3 introduced integrated software by adding graphics and database to the original spreadsheet concept. It is fast in operation since both the program itself and the current worksheet are held in the computer's Random Access Memory (RAM). It also includes its own programming (or, strictly speaking, macro) language with the ability to set-up user defined menus which operate just like the real thing.

Since 1-2-3 was launched it has been followed by software which integrates yet more functions, particularly word processing and communications with the original database, graphics and spreadsheet. This trend is typified by Symphony, Jazz and Framework. They have not, however, succeeded in toppling 1-2-3 from the top of the best seller lists. This is partly due to price but also a reflection of complexity. 1-2-3 has proved to have just the right trade-off between functionality (what it can do) and usability (how you go about it).

In fact, just as copies or 'clones' of the IBM PC have appeared, so too have lookalike versions of 1-2-3. Programs such as VP Planner and the Twin have functions and interfaces very similar to 1-2-3, differing only in detail or in extensions. Just like clone hardware, these software products are much cheaper than the real thing.

The following sections look at two clone spreadsheets, **As-Easy-As** a 'shareware' product and **Quattro** from Borland.

19.6 As-Easy-As

AS-EASY-AS is an example of shareware: that is it may be freely copied to other people for evaluation purposes, but if an individual wishes to continue using it they should register their copy by paying a fee. Registration brings a full manual and the possibility of future upgrades. It is a reasonable clone of Lotus 1-2-3 which can be customised further by typing [/] to call up the menu, **F**ile **R**etrieve **LOTUS**. The user should follow the instructions on the screen then type [/] **W**orksheet **E**rase **Y**es. Its main limitation is that the **D**ata **R**egression option only supports Linear Regression (Chapter 12) and not Multiple Regression (Chapter 18). It also omits **D**ata **D**istribution, but includes **A**rray (Data Matrix) **A**ddition, **S**ubtraction and solution of linear **E**quations. Other features include **R**ange **A**udit, **D**ata **G**oalseek, and **G**raph **P**lot It is the cheapest Lotus 1-2-3 clone on the market.

Table 19.7 As-Easy-As Command Equivalents

```
L O T U S   1 - 2 - 3              A S - E A S Y - A S
Command:  Meaning:                  Command:    Meaning:.
/Q        Quit                      /E    Exit
/RT       Range Transpose           /AT   Array Transpose
/DMI      Data Matrix Invert        /AI   Array Invert
/DMM      Data Matrix Multiply      /AM   Array Multiply
```

19.7 Quattro

Quattro implies "four" in succession to 1-2-3. It can be configured to be an exact copy of Lotus 1-2-3, except for minor features such as its reference to ranges as blocks. To do this, call up the menu and select **D**efault **S**tartup **M**enu Tree. As an Alternate Menu 123.ALT will be displayed. Select **S**witch Menus, **A**lternate Menu. Now when the [/] key is pressed again, a 1-2-3 style menu is obtained.

In addition to Lotus 1-2-3 combatibility Quattro offers **R**ange Search/replace, **F**ile **!**SQZ! compression, **G**raph **C**ustomisation and Printing, **G**raph **O**ptional Patterns and **M**arkers, **D**ata **S**ort with up to **5** keys and **D**ata **Q**uery **N**ames. It is probably the best Lotus 1-2-3 clone on the market.

19.8 Current Applications

VisiCalc was initially developed for financial calculations, and this is still the prime use of spreadsheets in business (Schuyler, 1984). However, spreadsheets have now spread into a variety of other application areas, particularly in education (Wozny, 1984). They have been widely used in mathematics where a good textbook is by Arganbright (1985). Other articles include Khaira (1986) and Soper & Lee (1985 & 1987), the last two of which cover means, standard deviations and regression.

Applications in engineering are covered by Haynes (1985) and Rao (1984), the first of which shows how to use 1-2-3 as a general purpose drawing tool. Applications in science include ecology (Silvert, 1984) and computer science (Shinners-Kennedy, 1986). Examples of spreadsheet work in psychology are by Hewett (1985) and Lee & Soper (1986), the last of which demonstrates the analysis of variance.

In the social sciences, an example in Geography is by Lee & Soper (1987) on the chi-squared test. Examples in Economics are the complicated macroeconomic model by Miller & Kelso (1985), and input-output analysis, index numbers and others by Judge (1989).

19.9 Future Applications

For the future, Kay (1984) sees the spreadsheet as an ultra-high level language whilst Cohen (1984) uses it as an ordinary programming language. Matheny (1984) sees its great potential for simulation studies, with such uses as compound interest, population explosion, radioactive decay and predator-prey interaction. In reality the spreadsheet is an electronic version of a blank sheet of paper. Just as the limits on what we can put down on a piece of paper are set by our own imagination, so too are the boundaries of the use of spreadsheets.

Bibliography

Arganbright, D. E. 1985 "Mathematical applications of electronic spreadsheets" McGraw-Hill. Includes statistics, finance and linear programming amongst others. Most suitable for mathematicians.

Berenson, M. & Levine, D. 1979 "Basic Business Statistics" Prentice Hall. Includes decision and time series analyses.

Chandy, P. R. & Garrison, S. 1985 "The use of spread sheets in the business/finance classroom" Collegiate Microcomputer **III**, 1, 69-74. Features VisiCalc.

Cohen, S. 1984 "Multiplan as a programming language" Drexel University microcomputing working paper series **F 84-3**. Shows how traditional algorithms can be translated into Multiplan.

Curwin, J. & Slater, R. 1985 "Quantitative Methods for Business Decisions" Van Nostrand Reinhold. Includes financial calculations.

Dobson, W. G. & Wolff, A. K. 1984 "Engineering problem solving with spreadsheet programs" Binary Engineering Associates. Includes stress analysis and heat transfer.

Haynes, J. L. 1985 "Circuit design with Lotus 1-2-3" Byte **10**, 11, 143-156. Shows how to use 1-2-3 as a general purpose drawing tool.

Hewett, T. T. 1985 "Teaching students to model neural circuits and neural networks using an electronic spreadsheet simulator" Behavior Research Methods, Instruments, & Computers **17**, 2, 339-344. Psychological application.

Hewett, T. T. 1986 "Using an electronic spreadsheet simulator to teach neural modeling of visual phenomena" Collegiate Microcomputer **IV**, 2, 141-151. Psychological application.

Hodge, S. E. & Seed, M. L. 1977 "Statistics and Probability" Blackie. Basic statistics textbook.

Hsiao, M. W. 1985 "A CAI framework of multiple regression analysis on spreadsheets" Collegiate Microcomputer **III**, 3, 239-247. Features 1-2-3.

Jackson, M. 1985 "Creative modelling with Lotus 1-2-3" Wiley. Includes floppy disk.

Judge, G. 1989 "Quantitative analysis for economics and business: the use of Lotus 1-2-3" Harvester Wheatsheaf.

Kay, A. 1984 "Computer software" Scientific American **251**, 3, 41-47. The spreadsheet as an ultra high level language!

Khaira, M. S. 1986 "Super class" Personal Computer World **9**, 3, 110-117. Aids to teaching maths and science.

Landram, F. G.; Cook, J. R. & Johnston, M. 1986 "Spreadsheet calculations of probabilities from the F, t, Chi-squared, and normal distribution" Communications of the ACM **29**, 11, 1090-1092. Uses a spreadsheet to automate the looking-up of statistical tables.

Lee, M. P. & Soper, J. B. 1986 "Using spreadsheets to teach statistics in psychology" Bulletin of the British Psychological Society **39**, 365-367. Illustrates the analysis of variance.

Lee, M. P. & Soper, J. B. 1987 "Using spreadsheets to teach statistics in geography" Journal of Geography in Higher Education **11**, 1, 27-33. Illustrates the chi-squared test.

Matheny, A. 1984 "Simulation with electronic spreadsheets" Byte **9**, 3, 411-414. Illustrates compound interest, population explosion, radioactive decay and predator-prey simulations.

Miller, R. M. & Kelso, A. S. 1985 "Analyzing government policies" Byte **10**, 10, 199-210. Illustrates the Klein macroeconomic model.

Rao, N. D. 1984 "Typical applications of microcomputer spreadsheets to electrical engineering problems" IEEE Transactions on Education **E-27**, 4, 237-242. Shows that spreadsheets are just as valuable in technological education.

Schuyler, M. 1985 "The evolution of spreadsheets" Microcomputers for information management **2**, 1, 11-23. A potted history.

Shinners-Kennedy, D. 1986 "Using spreadsheets to teach computer science" ACM SIGCSE Bulletin **18**, 1, 264-270. Computer science application.

Silvert, W. 1984 "Teaching ecological modelling with electronic spreadsheets" Collegiate Microcomputer II, 2, 129-133. Biological application.

Simkin, M. G. 1986 "Nonfinancial modeling techniques on electronic spreadsheets" Collegiate Microcomputer IV, 2, 165-171. Covers decision analysis.

Soper, J. B. & Lee, M. P. 1985 "Spreadsheets in teaching statistics" The Statistician 34, 317-321. Illustrates means and standard deviations.

Soper, J. B. & Lee, M. P. 1987 "Calculations and 'What if?' calculations with spreadsheets" Micromath 3, 1, 28-29. Illustrates regression.

Thomas, J. J. 1983 "An Introduction to Statistical Analysis for Economists" Weidenfeld & Nicolson. Includes multiple regression.

Williams, G. & Williams, D. 1985 "VisiCalc in linear algebra" Collegiate Microcomputer III, 1, 75-78. Covers matrix operations.

Wozny, L. 1984 "The spreadsheet in an educational setting" Drexel University microcomputing working paper series F 84-4. Includes finance and statistics.

Index

@AVG 49-50, 72
@COUNT 76
@DCOUNT 76, 85, 123
@EXP 96, 180
@FV 177
@IRR 171
@LN 180
@LOG 182, 184
@NPV 171
@PI 96
@PMT 175
@PV 176
@RAND 114-115
@SQRT 53
@STD 54, 72, 116-117
@SUM 49-50
@VLOOKUP ....
.... 91-92, 115-116

Absolute address 51-52
Absolute key 52
Access system 4
Addition rule 81-83
Additive model 160-163
Address, cell 47
Analysis of variance .. 147-153
And criterion 83-87
ANOVA 147-153
Argument 49
Arithmetic formula 37
Arithmetic mean 47-50
Arrow key 5
As-Easy-As 212
Average 47-50

Back out 15
Bar chart 18-19, 93-95
Bayes theorem 86
Between 147
Between SS 148-150, 152
Binomial coefficients 90-93
Binomial distribution 89-95

Calc key 84, 115, 117-118
Capital ....
.... 167-171, 173, 175-177
Cause 129
Cell address 38, 47
Cell pointer 4, 6-8
Chi-squared test 123, 125-127
Chi-squared values 101
Column labels 8
Column width 42, 45-46
Combinations 89
Command Menus 13-16
Commands 4
Compound interest ... 167-169
Conditional probability ....
.... 83-87, 106-107
Confidence limits ....
.... 116-118, 142-143
Constraint 187-191
Contingency table 122
Continuous distribution ....
.... 95-101
Continuous variable 88
Control panel 4
Copy command 13-15, 31
.... see also Range Value
Copy to 66
Copying commands 65

217

Copying formatted cells.....53
Copying formulae.................
.........................51-53, 63-66
Corrections..........................4
Correlation.......128, 136-137
Correlation, multiple............
.................192, 194, 196-198
Cost 170-172
Covariance........132, 199-202
Criterion - ..... see Data Query
Criterion range................ 123
Criterion, and.............. 83-84
Criterion, blank.................80
Criterion, or 81-83
Cross classification.......... 122
Crosstabulation................ 122
Cumulative frequency ....
..............................57, 64-66
Current cell........................4
Cursor.................................4
Cyclical pattern 159

Data disc.....................4, 225
Data distribution...........61-63
Data Fill 32-34, 61-62, 69
Data Matrix.........................
.... 193-195, 199-202
Data Query 77-85
Data Query Criterion ...77-85
Data Query Extract......78-85
Data Query Find 78
Data Query Input 77-78, 85
Data Query Output..78-82, 85
Data range 18, 21
Data Regression...................
.......................193-197, 203
Data Sort 34-36
Data sub-menu..................15
Data Table 123-125, 171
Data Table 1.................... 124
Data Table 2.................... 124
Database 76-87, 123
Decision analysis........103-109
Dependent variable .. 129, 192
Disc, data....................4, 225
Disc, system 4

Discount factor................170
Discount rate 169-172
Discrete variable...............88
Double-log.............. 183-184
Durbin-Watson statistic....199

EDIT...............................42
Edit key....................... 42, 84
Effect.............................129
End key 173
Enter............................ 6-7
Entry line......................... 4
Erase, range 58-60
Erase, worksheet......... 24, 42
Estimating......................140
Exit.................................11
Expected payoffs...... 103-106
Expected values 125
Explanation line.................. 4
Exponential growth.............
.... 168, 181-182
Exponential smoothing ....
.... 155, 158-159
Exponentiation 53
Extract - ....see Data Query

F ratio ....
... 148-149, 151-153, 197-198
F values 99-101
Factorial.....................89-93
Field names 77-80
File Save 10, 24
File sub-menu.........10, 14-15
Filename..........................10
Financial calculations 167-177
Find - ....see Data Query
Forecast................... 162-166
Forecasting 139
Format 42
Format Currency..............44
Format Date 44
Format Fixed...............44-45
Format General 43-44
Format , graph - ....
.... see Graph Options
Format message................45

Format Percent..................44
Format Reset......................44
Format Scientific...............44
Format sub-menu..............44
Format Text......................44
Format when copying........53
Format [,] 44
Formula 37-40
Formula into number ....
.................................84, 193
Frequency distribution ....
..............................57, 61-63
Function key 6
Functions, inbuilt..............49
Future value................... 177

Goto key......................... 173
Gradient........................ 131
Graph Color....................26
Graph command.......... 14-15
Graph Data-labels..............26
Graph Format 20, 25
Graph Grid 26
Graph Legend 22, 25
Graph, line............154-166
Graph Name................... 130
Graph Name Use.............. 138
Graph Options... 20-22, 25-26
Graph Reset 20
Graph Save 19
Graph Scale.....................26
Graph, storing................ 130
Graph sub-menu 17
Graph Titles.....................26
Graph Type.. 16-18, 20, 22-24
Graph, using stored......... 138
Graph View 18
Graphical solution.....188-191
Graphs of distributions 66

Help key 6
Home key..................... 173
Horizontal series...............15

Independent events............86
Independent variable 129,

... 192-193, 195-198, 200-201
Index numbers...........68, 213
Initial capital 167-169
Integrated software .. 205, 211
Intercept........................131
Interest rate 167-169, 173-177
Internal rate of return 170-171

Joint probability....83-85, 107

Keyboard........................ 5

LABEL 7, 9
Label prefix 29-31
Labels............................. 6
Labels as range names...........
...............................129, 133
Least squares................144
Letter, first 15
Likelihood 106
Line chart 24
Line graph 182
Linear programming 186-191
Lines, drawing..................31
Log-normal............ 179-180

Main menu 10, 13
Manual recalculation 115
Manual scaling, graphs.......98
Matrix...........................193
Matrix inversion..................
.... 193, 195, 201-202
Matrix multiplication...........
.....................193-195, 201-202
Matrix transposition.............
.... 193-194, 200
Maximum 28, 36-37
Mean 41, 47-50
Mean, frequency distribution.
..................................69-71
Mean sum of squares............
.... 148-153
Mean, weighted................68
Median 28, 35-37
Menu.................. 4, 10, 12-16
Minimum..............28, 36-37

219

Mistake.................... 7, 15
Move command.... 14-15, 113
Moving means 155-158
MultiPlan.................209-211
Multiple correlation..............
..................192, 194, 196-198
Multiple regression...192-202
Multiplication rule...............
.....................86-87, 107
Multiplicative model............
...............................163-166
Mutually exclusive events...81

Name Labels........... 129, 133
Name, range.......... 39, 49, 56
Name Use 138
Net present value 169-172
Normal density............ 95-98
Normal distribution 95-98

Objective function........... 187
Observed values.............. 125
Offset...............................85
One-way ANOVA......147-151
Optimal solution 187, 191
Or criterion 81-83
Outlier influence.......144-145
Output -.......... see Data Query
.... and Data Regression

Page down key................. 173
Page up key..................... 173
Payment...................173-177
Payoff-table..............103-104
Periodicity...............157-158
Pie chart 16-18, 23
POINT........................38, 46
Pointer... 4, 6-9, 38-39, 47-49
Pointer, menu...........10, 13
Pointing at cell addresses.......
...............................38-39, 47-49
Population 110, 114
Positioning labels -...............
..................see Label Prefix
Positioning values - ....
.... see Range Format

Positively skew 179
Posterior expected payoff......
.... 106-108
Posterior probability............
.... 86-87, 106-108
Prediction............... 137-139
Present value.... 169-172, 176
Principal................. 174-176
Print Align.......................73
Print Borders 73
Print Clear 74
Print command............14-15
Print Footer 73
Print Go...........................72
Print Header.....................73
Print Line 73
Print Margins 73
Print Options...................73
Print Other......................73
Print Page 73
Print Page-length..............73
Print Range 72
Print Setup.......................73
Print sub-menu 72
Printgraph 19
Probability..................75-87
Probability, conditional ....
.... 83-87, 106-107
Probability density............88
Probability distributions ....
...............................88-102
Probability, joint...83-85, 107
Probability, subjective ....
.... 105-106

Quartiles................28, 35-37
Quattro................... 212-213
Quit.................... 11, 19
Quit command.................15

R^2 .... 192, 194, 196-198
R^2, adjusted 198
Random sample...............110
Range.................... 13, 16, 29
Range, end 16
Range Erase 58-60

Range Format..................43
Range Label-prefix......29-31
Range Name..........39, 49, 56
Range Name Labels..129, 133
Range, start......................16
Range sub-menu..........13, 16
Range Transpose.................
..................193-194, 199-200
Range Value...................202
Rate 167-177
READY..................5, 7-9, 15
Recalculation...............54-55
Recalculation manual.......115
Record 77
Regression. 128-146, 192-204
Regression line..........131-145
Regression, multiple..192-204
Regular payments 173-177
Relative address............51-52
Relative frequency...............
.........................57, 63-64, 75
Release 2..................193-195
Residuals 131, 140,
..........143-144, 155, 197, 199
Residuals, sum of squares......
....................................140, 144
Return.....................169-172
Row labels 8

Sample means...........118-120
Sample replication 117
Sample standard deviation.....
.................................116-117
Save 10
Save, graph -........................
. see Graph Name, Graph Save
Save, spreadsheet - ....
........................see File Save
Scatter diagram ....
.......................128, 129-130
Screen 4
Seasonal analysis.......159-166
Semi-log..................181-182
Slope...............................131
Sort Data-range 34
Sort Primary-key..............34

Sort Secondary-key...........35
Stacked-Bar chart..............23
Standard deviation.............
......................41, 50-51, 71
Standard deviation, sample....
.................................116-117
Standard error. 140, 192-193,
................196-197, 200-202
Standard error of a...140-141
Standard error of b...140-141
Standard error of prediction..
.................................142-143
Standard error of the mean
.................................117
Standardised normal..........98
Stationary series..............155
Statistical inference 110
Status line........................4
Sub-menu........................15
SuperCalc 207-209
Symbols...........................21
System disc......................4

t values ....
.........101, 141-143, 196-197
Table key 125
Time series.............154-166
Title label........................7
Total SS 149-150, 152
Transformation ....
................128, 134, 178-185
Trend....................155-156
Two-way ANOVA...151-153
Two-way table................122

VALUE..................9, 40, 47
Value............................7
Value, characters beginning...
................................47, 56
Variable..........................9
Variance ....
............53, 71, 132, 199-202
VisiCalc.................205-207

Within.......................148
Within SS.........148-150, 152

221

Worksheet Column-width .. 46
Worksheet Delete.......... 58-59
Worksheet Erase........... 24, 42
Worksheet Format 43, 45
Worksheet Insert 69
Worksheet Label-prefix..... 30
Worksheet sub-menu 13, 15

X range 18, 20
XY graph......... 20, 24, 97-98,
.... 129-130, 134, 138-139,
.... 142-143, 179-185

Z values 101

[#] in header 73
[$] 51
[(] 54
[*] 38
[+] 47
[,] 44
[.] 16, 18
[/] 10-11, 13, 38, 56
[@] 48
[@] in header 73

[^] 53
[|] 73
[1/HELP] 6
[2/EDIT] 42, 84
[4/ABS] 52
[5/GOTO] 173
[8/TABLE] 125
[9/CALC] 84, 115, 117-118
[BACKSPACE].................
.... 4, 6-7, 16, 36, 42
[BREAK] 16
[CAPS] 5-7
[DOWN] 6-8, 18
[END] 173
[ENTER] 4
[ESC] 6, 15, 24
[HOME] 173
[LEFT] 6, 8, 15, 42, 46
[PGDN] 173
[PGUP] 173
[RETURN] ....
... 4, 6-9, 15-18, 39, 42, 46, 49
[RIGHT] 6-8, 13-15, 42, 46
[SHIFT] 5, 7
[SPACE] 8
[UP] 6, 8

The Authors

Jean Soper is a Lecturer in Economics at the University of Leicester and Martin Lee is a Senior Lecturer in Computing Science at the Royal Military College of Science, Shrivenham. They have developed the use of spreadsheets in teaching Quantitative Techniques across a wide range of disciplines including Statistics, Economics, Psychology, Geography, Mathematics and Computing. Students on the courses they have taught together have included both undergraduates and Businessmen. They have published four academic papers stemming from their work.

This, their first book, has grown out of their teaching and now provides a crystallisation of their best ideas.

Jean B. Soper

Born in Sheffield on 19th September, 1946, Jean Soper was educated at Dollar Academy and Dundee High School, where she was dux in Mathematics. In 1967 she graduated from the University of St. Andrews with a joint honours degree in Economics and Statistics, and was awarded the N.C.R. prize in Statistics. She joined the staff at the University of Leicester as a Research Assistant on the East Midlands Retail Trade Project, and was appointed Lecturer in 1969. Her M.Phil., completed in 1973, used statistical techniques to draw up a socio-economic classification of East Midlands Towns. She is a Fellow of the Royal Statistical Society.

All specialist Economists at Leicester University are taught to use spreadsheets by Jean and her colleagues. Jean also teaches Statistics and Economics, and is an Admissions Tutor and Schools Relations Officer.

Jean has applied quantitative techniques in various fields making considerable use of computers. She has contributed to a number of statistical, social science and medical journals. A project commissioned by Leicester City Council to forecast economic trends in Leicester has recently been completed.

Jean is married with three school age sons. The family live in an East Leicestershire conservation village. They enjoy relaxing in their large garden, participating in village activities, and driving across Europe on camping holidays.

Martin P. Lee

Martin Lee was born on 2nd September, 1953, in Harlow, Essex, where he lived for his first 19 years. Educated at Burnt Mill Comprehensive School, he was a member of the BBC TV "Top of the Form" team which reached the final. In 1975 he gained a First Class B.Sc. degree in Geography and Geology at King's College London where he was Tennant Exhibitioner in Geology, Sambrooke Exhibitioner in Natural Sciences and Wooldridge Prizewinner in Geography. He was later awarded the degree of M.Sc. by the University of Keele for a thesis entitled "Some properties of recent tills from Central Norway".

In 1978 he became a teaching fellow at the University of Lancaster and also began teaching for the Open University. From 1981 to 1987 he was a Lecturer in the Department of Computing Studies at the University of Leicester. Since 1987 he has been a Senior Lecturer in the Computing Science Group at the Royal Military College of Science, which is part of the Cranfield Institute of Technology. He teaches databases, data administration and Prolog on the BSc degree in Command & Control, Communcations & Information Systems (CIS) and is academic adviser to the MSc degree in Design of Information Systems (DIS) at the college. He also teaches Management Information Systems (MIS) on the Master of Defence Administration (MDA) degree and databases on the MSc in Military Electronic Systems Engineering (MESE) .

Martin has published over twenty-five papers on a wide variety of topics including spreadsheets, database, computer graphics, operating systems, human-computer interfaces, simulation, statistics, knowledge-based systems and fourth & fifth generation programming languages. He is a Member of the British Computer Society.

His hobbies include walking his Labrador dog Jasper on the Berkshire Downs, boating on the Thames, caravanning across Britain, and gardening at his home in Didcot, Oxfordshire.

Floppy discs

All the completed spreadsheets in this book are available for IBM PC compatible machines. They come on either two 360kb format 5 1/4" floppy discs or one 720kb format 3 1/2" floppy disc together with an evaluation copy of As-Easy-As, the best value 1-2-3 'clone', all for £10.00. The first disc will enable you to load and inspect completed versions of all the examples in the book and to evaluate more easily the computing techniques we are using. The second disc allows you to perform nearly all the procedures given in our book without having access to Lotus 1-2-3 .

Order Form

To: Mach One Computing Limited
      1 Loncot Road
      Shrivenham
      SWINDON
      Wiltshire
      SN6 8AL

Please find enclosed a cheque for £10.00 made payable to "Mach One Computing Limited" for:

either 2 x 5 1/4" **or** 1 x 3 1/2" (please delete)

floppy disc(s) containing all the completed spreadsheets in this book together with an evaluation copy of As-Easy-As.

From: ....(Name)
      ....(Address)
      ....(Town)
      ....(County)
      ....(Postcode)
      .... (Country)

Computing Books from Chartwell-Bratt

GENERAL COMPUTING BOOKS

Compiler Physiology for Beginners, M Farmer, 279pp, ISBN 0-86238-064-2
Dictionary of Computer and Information Technology, D Lynch, 225 pages, ISBN 0-86238-128-2
File Structure and Design, M Cunningham, 211pp, ISBN 0-86238-065-0
Information Technology Dictionary of Acronyms and Abbreviations, D Lynch, 270pp, ISBN 0-86238-153-3
The IBM Personal Computer with BASIC and PC-DOS, B Kynning, 320pp, ISBN 0-86238-080-4

PROGRAMMING LANGUAGES

An Intro to LISP, P Smith, 130pp, ISBN 0-86238-187-8
An Intro to OCCAM 2 Programming, Bowler, et al, 109pp, ISBN 0-86238-137-1
Cobol for Mainframe and Micro: 2nd Ed, D Watson, 177pp, ISBN 0-86238-211-4
Comparative Languages: 2nd Ed, J R Malone, 125pp, ISBN 0-86238-123-1
Fortran 77 for Non-Scientists, P Adman, 109pp, ISBN 0-86238-074-X
Fortran 77 Solutions to Non-Scientific Problems, P Adman, 150pp, ISBN 0-86238-087-1
Fortran Lectures at Oxford, F Pettit, 135pp, ISBN 0-86238-122-3
LISP: From Foundations to Applications, G Doukidis et al, 228pp, ISBN 0-86238-191-6
Prolog versus You, A Johansson, et al, 308pp, ISBN 0-86238-174-6
Simula Begin, G M Birtwistle, et al, 391pp, ISBN 0-86238-009-X
The Intensive C Course: 2nd Edition, M Farmer, 186pp, ISBN 0-86238-190-8
The Intensive Pascal Course: 2nd Edition, M Farmer, 125pp, ISBN 0-86238-219-X

ASSEMBLY LANGUAGE PROGRAMMING

Coding the 68000, N Hellawell, 214pp, ISBN 0-86238-180-0
Computer Organisation and Assembly Language Programming, L Ohlsson & P Stenstrom, 128pp, ISBN 0-86238-129-0
What is machine code and what can you do with it? N Hellawell, 104pp, ISBN 0-86238-132-0

PROGRAMMING TECHNIQUES

Discrete-events simulations models in PASCAL/MT+ on a microcomputer, L P Jennergren, 135pp, ISBN 0-86238-053-7
Information and Coding, J A Llewellyn, 152pp, ISBN 0-86238-099-5
JSP - A Practical Method of Program Design, L Ingevaldsson, 204pp, ISBN 0-86238-107-X

HARDWARE

Computers from First Principles, M Brown, 128pp, ISBN 0-86238-027-8
Fundamentals of Microprocessor Systems, P Witting, 525pp, ISBN 0-86238-030-8

NETWORKS

Communication Network Protocols: 2nd Ed, B Marsden, 345pp, ISBN 0-86238-106-1
Computer Networks: Fundamentals and Practice, M D Bacon et al, 109pp, ISBN 0-86238-028-6
Datacommunication: Data Networks, Protocols and Design, L Ewald & E Westman, 350pp, ISBN 0-86238-092-8
Data Networks 1, Ericsson & Televerket, 250pp, ISBN 0-86238-193-2
Telecommunications: Telephone Networks 1, Ericsson & Televerket, 147pp, ISBN 0-86238-093-6
Telecommunications: Telephone Networks 2, Ericsson & Televerket, 176pp, ISBN 0-86238-113-4

GRAPHICS

An Introductory Course in Computer Graphics, R Kingslake, 146pp, ISBN 0-86238-073-1
Techniques of Interactive Computer Graphics, A Boyd, 242pp, ISBN 0-86238-024-3
Two-dimensional Computer Graphics, S Laflin, 85pp, ISBN 0-86238-127-4

APPLICATIONS

Computers in Health and Fitness, J Abas, 106pp, ISBN 0-86238-155-X
Developing Expert Systems, G Doukidis, E Whitley, ISBN 0-86238-196-7
Expert Systems Introduced, D Daly, 180pp, ISBN 0-86238-185-1
Handbook of Finite Element Software, J Mackerle & B Fredriksson, approx 1000pp, ISBN 0-86238-135-5
Inside **Data Processing: computers and their effective use in business,** A deWatteville, 150pp, ISBN 0-86238-181-9
Proceedings of the Third Scandinavian Conference on Image Analysis, P Johansen & P Becker (eds) 426pp, ISBN 0-86238-039-1
Programmable Control Systems, G Johannesson, 136pp, ISBN 0-86238-046-4
Risk and Reliability Appraisal on Microcomputers, G Singh, with G Kiangi, 142pp, ISBN 0-86238-159-2
Statistics with Lotus 1-2-3, M Lee & J Soper, 207pp, ISBN 0-86238-131-2

HCI

Human/Computer Interaction: from voltage to knowledge, J Kirakowski, 250pp, ISBN 0-86238-179-7
Information Ergonomics, T Ivegard, 228pp, ISBN 0-86238-032-4
Computer Display Designer's Handbook, E Wagner, approx 300pp, ISBN 0-86238-171-1

Linear Programming: A Computational Approach: 2nd Ed, K K Lau, 150pp, ISBN 0-86238-182-7
Programming for Beginners: the structured way, D Bell & P Scott, 178pp, ISBN 0-86238-130-4
Software Engineering for Students, M Coleman & S Pratt, 195pp, ISBN 0-86238-115-0
Software Taming with Dimensional Design, M Coleman & S Pratt, 164pp, ISBN 0-86238-142-8
Systems Programming with JSP, B Sanden, 186pp, ISBN 0-86238-054-5

MATHEMATICS AND COMPUTING

Fourier Transforms in Action, F Pettit, 133pp, ISBN 0-86238-088-X
Generalised Coordinates, L G Chambers, 90pp, ISBN 0-86238-079-0
Statistics and Operations Research, I P Schagen, 300pp, ISBN 0-86238-077-4
Teaching of Modern Engineering Mathematics, L Rade (ed), 225pp, ISBN 0-86238-173-8
Teaching of Statistics in the Computer Age, L Rade (ed), 248pp, ISBN 0-86238-090-1
The Essentials of Numerical Computation, M Bartholomew-Biggs, 241pp, ISBN 0-86238-029-4

DATABASES AND MODELLING

An Introduction to Data Structures, B Boffey, D Yates, 250pp, ISBN 0-86238-076-6
Database Analysis and Design, H Robinson, 378pp, ISBN 0-86238-018-9
Databases and Database Systems, E Oxborrow, 256pp, ISBN 0-86238-091-X
Data Bases and Data Models, B Sundgren, 134pp, ISBN 0-86238-031-6
Text Retrieval and Document Databases, J Ashford & P Willett, 125pp, ISBN 0-86238-204-1
Towards Transparent Databases, G Sandstrom, 192pp, ISBN 0-86238-095-2
Information Modelling, J Bubenko (ed), 687pp, ISBN 0-86238-006-5

UNIX

An Intro to the Unix Operating System, C Duffy, 152pp, ISBN 0-86238-143-6
Operating Systems through Unix, G Emery, 96pp, ISBN 0-86238-086-3

SYSTEMS ANALYSIS AND DEVELOPMENT

Systems Analysis and Development: 3rd Ed, P Layzell & P Loucopoulos, 284pp, ISBN 0-86238-215-7

SYSTEMS DESIGN

Computer Systems: Where Hardware meets Software, C Machin, 200pp, ISBN 0-86238-075-8
Distributed Applications and Online Dialogues: a design method for application systems, A Rasmussen, 271pp, ISBN 0-86238-105-3

INFORMATION AND SOCIETY

Access to Government Records: International Perspectives and Trends, T Riley, 112pp, ISBN 0-86238-119-3
CAL/CBT - the great debate, D Marshall, 300pp, ISBN 0-86238-144-4
Economic and Trade-Related Aspects of Transborder Dataflow, R Wellington-Brown, 93pp, ISBN 0-86238-110-X
Information Technology and a New International Order, J Becker, 141pp, ISBN 0-86238-043-X
People or Computers: Three Ways of Looking at Information Systems, M Nurminen, 1218pp, ISBN 0-86238-184-3
Transnational Data Flows in the Information Age, C Hamelink, 115pp, ISBN 0-86238-042-1

SCIENCE HANDBOOKS

Alpha Maths Handbook, L Rade, 199pp, ISBN 0-86238-036-7
Beta Maths Handbook, L Rade, 425pp, ISBN 0-86238-140-1
Handbook of Electronics, J de Sousa Pires, approx 750pp, ISBN 0-86238-061-8
Nuclear Analytical Chemistry, D Brune *et al,* 557pp, ISBN 0-86238-047-2
Physics Handbook, C Nordling & J Osterman, 430pp, ISBN 0-86238-037-5
The V-Belt Handbook, H Palmgren, 287pp, ISBN 0-86238-111-8

Chartwell-Bratt specialise in excellent books at affordable prices.

For further details contact your local bookshop, or ring Chartwell-Bratt direct on **01-467 1956** (Access/Visa welcome.)

Ring or write for our *free* catalogue.

Chartwell-Bratt (Publishing & Training) Ltd, Old Orchard, Bickley Road, Bromley, Kent, BR1 2NE, United Kingdom.
Tel 01-467 1956, Fax 01-467 1754, Telecom Gold 84:KJM001,
Telex 9312100451(CB)

Teaching Statistics in the Computer Age
Lennart Råde (editor)

Tackling the implementation and usage of computers for the teaching of statistics, at university and school level, this book offers the combined experience of seventeen experts from different parts of the world. With thorough coverage of a wide range of aspects it specifically deals with the introduction of computer-intensive methods in university courses. University departments report on their use of mainframes and micros and the use of simulation, statistical data analysis, computer animation and programmable calculators are discussed in depth.

Contents: The impact of calculators and computers on teaching statistics; Mastering elementary probability and calculator programming in a class; The place of computers in the learning of statistics; Are statistical tables obsolete? How calculators and computors change the field of problems in teaching statistics; Statistics and computer science - an integrated high school course; Using microcomputers to extend and supplement existing material for teaching statistics; Using microcomputers for data analysis and simulation experiments in junior and senior high school; Computer animation - a powerful way of teaching concepts of probability and statistics; Using microcomputers for teaching introductory statistics - experimental results and implications; New and improved statistical skills in the computer era; The microcomputer as an aid to instruction-problems and pitfalls; The pocket computer in the classroom; Use of computer in statistics courses at the University of Melbourne; Teaching statistics at the university level - how computers can help us find realistic models for real data and reasonably assess their reliability; The Nightingale data library and protocal - a proposal for a library of interesting data; The teaching of statistics in New Zealand schools; Bibliography; List of participants.

Hardback, 1985, 248 pages, ISBN 0-86238-090-1

phone 01-467 1956 for further details and to place your order